Primordial Prescription:

The Most Plaguing Problem of Life Origin Science

Dr. Abel is a life-origin specialist with scores of peer-reviewed science journal publications. He is the Editor of **The First Gene: The Birth of Programming, Messaging and Formal Control.** Dr. Abel is the Director of the Gene Emergence Project of the Origin of Life Science Foundation, Inc. His specialties are Proto-Biocybernetics and Proto-Biosemiotics.

Primordial Prescription:

The Most Plaguing Problem of Life Origin Science

David L. Abel

Director, The Gene Emergence Project,
Department of ProtoBioCybernetics and ProtoBioSemiotics,
The Origin of Life Science Foundation, Inc.
LongView Press–Academic New York, N.Y.

Copyright © 2015 by David L. Abel. All rights reserved. No part of this publication may be reproduced, stored in a retrieval system, or transmitted, in any form or by any means, electronic, mechanical, photocopying, recording, or otherwise, without the prior permission of LongView Press™ or the author, said permission being obtained from the writer. Contact Info: David L. Abel, Program Director, The Gene Emergence Project™: An international consortium of scientists investigating protocell control mechanisms. Department of ProtoBioCybernetics/ ProtoBioSemiotics, The Origin of Life Science Foundation, Inc. www.lifeorigin.org / email: life@us.net

Author: David L. Abel
Cover Illustration, design © 2015:   Jessie Nilo
Proofreader   Kathy Phipps
Interior photos, tables, charts and graphics: As noted in book

Library of Congress Control Number: 2015932686

Library of Congress Subject Heading Suggestions:

1. QH325   Life—Origin, Origin of beginning of life
2. sh85022912   Molecular Evolution
3. sh85126861   Spontaneous—Generation
4. QH508   Biological Control Systems
5. Q336   Cybernetics, Self-organizing systems. Conscious automata. Data processing.

BISAC Classification Suggestions:

1. SCI029000   SCIENCE / Life Sciences / Genetics & Genomics
2. SCI049000   SCIENCE / Life Sciences / Molecular Biology
3. SCI064000   SCIENCE / System Theory

ISBN-Print Edition: **978-0-9657988-2-2**   V 1.0   ISBN-eBook Edition: **978-0-9657988-3-9**

Abel, David L. 1946 —

Primordial Prescription: the Most Plaguing Problem of Life Origin Science
David L. Abel. Includes bibliographic references and index

LongView Press™—Academic, 2015
244 5th Avenue, Suite # G228, New York, NY 10001-7604
Printed in the United States of America on acid-free paper

Dedication

To Morris Wayne Hedge,

A lifelong friend, best man,

mathematician, scholar,

editor, patriot and

Chairman of the Board of the

Origin of Life Science Foundation, Inc.

Table of Contents

Title page   i
Copyright page   ii
Dedication   iii
Table of Contents   v
Preface   xi
Acknowledgements   xiii
Introduction   xv

Section	**Page Number**
1. The Most Plaguing Problem of Life Origin Science	1
It's not about information. It's about *Prescription, and its Processing*.	1
1.1. What exactly is prescription?	1
1.2. Prescription of what?	2
1.3. Information itself doesn't *do* anything.	4
1.4. Information discussions only lead to a quagmire.	5
2. What does Prescription require?	11
2.1. Prescription, unless we pioneer it, has to be instructed with prior purposeful choices	12
2.2. Prescription includes not only Instructions (Prescriptive Information [PI]), but the processing of those instructions.	13
2.3. The processing of Prescriptive Information (PI) itself requires PI.	14
2.4. Prescription comes in many forms	15
2.4.1. Instructions/directions (typically using semantic language expressed using a formal symbol system.)	15
2.4.1.1 Non-material Symbol Systems	15
2.4.1.2 Material Symbol Systems	16
2.4.2 Programming (e.g., the theoretical Turing tape)	18
2.4.3 Pre-set configurable switch-settings in an integrated circuit (e.g., chips, circuit boards)	21

	2.4.4	The integration/organization of component parts into a holistic Sustained Functional System (SFS)	23
	2.4.5	The design/engineering/manufacturing of those individual component parts needed to construct any non-trivial machine.	24
	2.4.6	Pre-programmed degrees of end-user freedom of choice	25

2.5 Prescription and its processing invariably pursues the *goal* of function ... 25
2.6 Prescription can only be verified through successful realization of utility ... 26
2.7 Prescription requires contingency. But, there are two kinds of contingency [The Universal Contingency Dichotomy (UCD)]: ... 26
    2.7.1 Chance contingency ... 27
    2.7.2 Choice contingency:
    Choice-Contingent Causation and Control (CCCC) ... 27
2.8 The Prescription Principle: Prescription requires Choice Contingency ... 29
2.9 The first step towards prescribing utility is usually "organization." ... 32
2.10 Organization requires choice contingency at decision nodes ... 33
2.11 Prescription is a function of Decision Theory, not Stochastic Theory. ... 33
2.12 Formalism vs. Physicality. What exactly is "formalism?" ... 34
    2.12.1 Physicodynamic Determinism (PD) ... 35
    2.12.2 Formalism and its inherent Choice Determinism (CD) ... 35
2.13 Prescription always emanates from the far formal side of the great divide known as The Cybernetic Cut ... 38
2.14 Prescription always flows across on the one-way Configurable Switch (CS) Bridge from formalism to physicality. ... 40
2.15 The fundamental unit of Prescriptive Choice (PC) is the binary "decision node." ... 44
2.16 Can programming choices be measured with fixed units of measurement? ... 46
2.17 Both PI, and the processing of PI, require Mind and Agency ... 50
    2.17.1 Self-awareness ... 53
    2.17.2 Valuation, Desire and Motivation ... 54
    2.17.3 Intention and the Pursuit of functionality ... 54
    2.17.4 The choice contingency needed to achieve formal utility ... 55
    2.17.5 Knowledge of/about ontologic being (objective reality) ... 55

3. How did primordial prescription and its processing arise in a prebiotic environment? ... 57

3.1 The chance and necessity of nature are blind to function and "usefulness." ... 58
3.2 No yet-to-be discovered law could possibly prescribe sophisticated function. ... 63
3.3 The Universal Selection Dichotomy (USD) ... 64
    3.3.1 Selection FROM AMONG ... 65

	3.3.2	Selection FOR (in pursuit of)	65
3.4	Natural Selection		70
	3.4.1	Which of the two categories of Selection might have been exercised by molecular evolution?	75
	3.4.2	Which of the two categories of Selection would have been exercised by post-biotic evolution?	76
	3.4.3	The GS Principle (The Genetic Selection Principle)	77
3.5	Molecular evolution offers only two possible kinds of "fitness"		79
	3.5.1	Greater molecular stability	79
	3.5.2	Mutual or self-replication	81
3.6	Selection Pressure		86
3.7	Directed Evolution		89
3.8	Evolutionary algorithms		90
3.9	Ontological vs. Epistemological Prescription and Processing (P & P)		92
3.10	Can Ontological Prescription, and its processing increase in nature?		95
3.11	Chaos theory's self-ordering cannot produce formal organization or nontrivial function.		96
	3.11.1	"Dissipative structures" dissipate!	101
	3.11.2	The relation of thermodynamics and statistical mechanics to ontologic prescription and its processing.	101
	3.11.3	A cell is a Sustained Functional System (SFS)	105
	3.11.4	A thermodynamically open environment is not sufficient to generate an SFS	106
	3.11.5	Entropy is better understood as disorganization than disorder.	110
3.12	Constraints vs. Controls		114
3.13	Laws vs. Rules		115
3.14	The Universal Determinism Dichotomy (UDD): Physicodynamic Determinism vs. Choice Determinism		120
	3.14.1	Physicodynamic Determinism (PD)	120
	3.14.2	Choice Determinism (CD)	120
3.15	Complexity does not equal abstract concept, organization or functionality.		122
3.16	The Formalism > Physicality Princple (F > P Principle)		123
3.17	No targets existed in a prebiotic environment, and no searches were conducted		128
3.18	Intuitive work vs. the physics definition of work.		131
	3.18.1	Physics definition of work. A tumble weed blown up hill is not intuitive, useful work	131
	3.18.2	Intuitive, useful work	132
3.19	Machines do intuitive, useful work		134
	3.19.1	How do machines and computers come into existence?	135

3.19.2	The simplest known cells require millions of molecular machines and nanocomputers	135
3.19.3	How did so many subcellular molecular machines come into existence?	137
3.19.4	Spontaneous generation of a simple paper clip	138
3.20 Probability vs. Plausibility		146
3.21 Bioinformational Turing tapes and machines		147
3.21.1	Software prescribes not-yet-existent formal function into existence	150
3.21.2	Venter's "Programming of a digital organism."	152
3.21.3	Prescription, with its programming and processing, are the most defining characteristics of life	158
3.22 Life is a programmed, cybernetic, highly-regulated, computational process		159
3.23 What could possibly produce subcellular "information technology"?		161
3.24 Duplication plus variation. Mutations result in noise, not increased P & P		164
3.25 Emergence		171
3.25.1	Hypercycles	180
3.25.2	Trial and Error Pursuits	182
3.25.3	Oscillation models of life origin	182
3.26 Definition of life		186
3.27 Challenges remaining in astrobiological research		190
3.27.1	The quandary of life origin is not complexity; it is *conceptual* complexity.	192
3.27.2	Multiple layers and dimensions of Prescription are superimposed in genomes	196
3.27.3	Queries needing answers for *any* naturalist model of life origin to be plausible	205
3.28 Primordial life would have required Prescription and its Processing.		210
3.28.1	Prescription and its Processing (P & P) are life's most essential ingredient	210
3.28.2	How does Choice Determinism relate to the spontaneous generation of life?	213
3.28.3	Biological PI_o is nonphysical, the same as any other PI	215
3.28.4	Life-Origin is about the emergence of Primordial P & P	219
3.28.5	The worship of "possibility"	224
3.28.6	Still another chicken and egg paradox to life origin	226
3.28.7	The multi-dimensional, multi-layered nature of biological prescription	228
3.28.8	The first P & P's were ontological, not epistemological	229
3.28.9	We must first address the origin of PI_o, before addressing the modification of existing PI_o	232

4. Summary	**239**
5. References	**267**
6. Index	**299**

Preface

Which came first, the chicken or the egg? *It doesn't matter.* Both the chicken and the egg are highly prescribed. Both are replete with extraordinary formal organization (not just the physicodynamic self-ordering that we see in chaos theory), purposefully set configurable switches, integrated circuits, computationally successful schemes, developmental plans, end-user programmable preferences, and contingency modules that prescribe appropriate responses to almost any environmental stress or opportunity. All of this genomic and epigenomic programming predates and "makes happen" phenotypic cells. This programming and processing alone compute and manufacture organisms with superior fitness. Environmental selection is secondary—determined only after the fact of this programming and processing at the decision-node level of genomic and epigenomic prescription.

Life pursues the *goal* of being and staying alive. Evolution has no goal, especially at the molecular/genetic/epigenetic programming level. Evolution is nothing more than the differential survival and reproduction of the fittest already-programmed, already-living organisms. Evolution only eliminates inferior phenotypic organisms. Evolution cannot prescribe or program.

All known life is cybernetic, meaning controlled. Life's most prominent attribute is programming and tight regulation at every turn. Yet programming, prescription, control and regulation are all formalisms, not mere physical interactions. The formal programming of life is what makes life unique. Heritable formal instructions are needed, independent of the actual phenotypic organisms, for evolution to be possible.

Metabolism-First models cannot sustain themselves as perpetual motion machines. It is not sufficient for an environment to be open to wasted solar energy. Sustained Functional Systems (SFS) are needed to convert wasted solar energy into usable energy potentials. Engines are needed to be able to use these energy potentials in an organized fashion. Selection must be made *FOR* (in pursuit of), not just Selection FROM AMONG in order to prescribe and process future function. Natural Selection is Selection FROM AMONG only. Evolution *never* Selects *FOR* (in pursuit of).

No reason exists why a prebiotic environment would have "cared" whether anything functioned. A basis for Selection *FOR* (in pursuit of) is completely lacking in any naturalistic abiogenesis theory. Yet programming and prescribing is impossible without selecting FOR (in pursuit of). This kind of selection is always formal, not physical.

Few answers have been published to any of the major questions posed in or by the Origin of Life Prize program, or by *The First Gene: the Birth of Programming, Messaging and Formal Control,* David L. Abel, Editor, LongView Press Academic, New York, *N.Y., 2011.*

Prescription and programming arise only out of Decision Theory, not Stochastic Theory. How did prebiotic nature program the first decision nodes? Only Choice Determinism (CD), not Physicodynamic Determinism (PD), could possibly program a genome and epigenome.

What does naturalistic science do with the fact that the phenomena of Prescription and its Processing have never once been observed independent of agent causation? That is the topic of this book.

David L. Abel, Director
The Gene Emergence Project
Department of ProtoBioCybernetics & ProtoBioSemiotics
The Origin of Life Science Foundation, Inc.

Acknowledgements

I would like to thank Morris W. Hedge for editing and proof-reading this volume.

Greg McElveen once again was a joy to work with. Greg did a marvelous job at LongView Press Academic in preparing the manuscript for publication, and in overseeing its processing.

Jessie Nilo did a very nice job on cover design and graphic art and Kathy Phipps did a great job on technical editing.

Finally, the author would like to acknowledge the financial support for this work coming in the form of a grant from The Origin of Life Science Foundation, Inc. An anonymous multi-millionaire donor has generously provided the funding over many years for every Gene-Emergence Project venture the Board of Directors has ever approved. The project has always focused research attention specifically on the problem of gene emergence, code-origin, and the beginning of epigenetic management. Of special interest has been the birth of programming, messaging, formal controls, and processing in a prebiotic naturalistic world.

Introduction

Life origin is *not* about information!

In addition, no searches were conducted in a prebiotic environment. No "targets" existed at which to aim; shooters were non-existent. No goals existed—no pursuits of formal organization or functionality.

Life origin is about initial *Prescription,* and about the *Processing* of that initial prescription. We are not talking about epistemology here. Epistemology pertains to the nature, scope and limits of human knowledge. Human knowledge did not exist when life began. We want to imagine how stand-alone, Ontological Prescription and its Processing (P & P_o) arose in a prebiotic environment. This has nothing to do with human uncertainty, reduced uncertainty, or "mutual entropy."

This book will demonstrate from many different angles that only P & P_o could have generated the first formal organization, sophisticated function, and the first non-trivial *useful* work.

Life exhibits the highest-tech organization and integrative function known to science. The simplest-known living organism puts to shame the finest mainframe computer system in the world. Life is highly controlled, regulated, and computational. Life is fundamentally cybernetic.

How P & P_o arose in inanimate nature is by far the most perplexing problem facing life-origin science. Some have argued that programming had to precede, organize, instruct, and "compute" life into existence. Others, realizing that prebiotic nature would have been incapable of practicing such formalisms, instead believe primordial cells "self-organized" spontaneously, and just "emerged" due to purely physical interactions.

Holders of both perspectives need desperately to critique this book. Can you vigorously defend your perspective under extensive challenge? Or is your mind fanatically "made up" prior to any substantive discussion?

Unfortunately, many will *not* accept the challenge. They will instead cowardly post a "one star" review of this book without reading past the first page. While boasting of academic superiority, their closed minds are in fact locked shut. They are "true believers" in superstition, tragically, all in the name of science.

Instead of sweeping the issue of Primordial Prescription and its Processing under the rug, this entire book focuses directly on it. The highly specialized field of science that specifically addresses these questions is known as ProtoBioCybernetics. Cybernetics is the study of control and regulation. "Bio," of course, refers to life. "Proto" addresses the emergence of initial life. Protocells would have been the very first semblance of living cells in any naturalistic scenario.

If biologists know anything at all about life, they know that every activity within even the simplest known living cell is exquisitely organized. Life is genomically and epigenomically controlled and regulated. How did that organization, prescription, processing, control and regulation of biofunction get started in inanimate nature? That is the subject of this book.

1. The Most Plaguing Problem of Life Origin Science

1.1 What exactly is Prescription?

Before we can talk about life, we have to clarify what is Prescription in everyday terms. What does it do? How is Prescription achieved? What is its nature? How is it recorded? How is it processed and employed? What are its capabilities?

All causation can be subdivided into one of these two categories: Physicodynamic causation (physical causes acting upon starting-condition constraints) and Formal causation (abstract, conceptual, non-physical, purposeful, choice-based decisions).

Prescription has nothing to do with human uncertainty, reduced uncertainty (mutual entropy), statistical combinatorialism, transmission theory, description, or human knowledge. Prescription is more than just instruction. Prescription cannot be reduced to just enumerating how something could, or should, be done. Successful prescription, as used with reference to abiogenesis (initial life origin), actually makes integrated metabolic function happen in a physical world. Prescription not only precedes in time what it prescribes, it plays the primary role of actually bringing what is prescribed into existence.

Prescription and its processing go hand in hand. Instructions are worthless if they cannot be processed. The value of a computational program requires processing to demonstrate. We cannot even be sure if a computational program will "halt" (successfully compute) without running it (the "halting problem" made famous by Alan Turing's proof of "undecidability").[1] The program must be processed to its completion to prove that its prescription is efficacious.

Causation, therefore, is the biggest issue of life origin science, not just information. Effects must be caused. Successful computation is an effect that must be programmed into existence. Programming, and its processing, is the cause. Since all known life is cybernetic (controlled and regulated), life-origin science must seek to explain how the first primordial cell (the effect) was caused (programmed and processed into existence).

Successful prescription of formal utility (usefulness) *executes* Prescriptive Information (PI).[2] Prescriptive Information is a certain kind of Functional Information (FI).[3-7] We will address the different types of true "information" later. The birth of Prescription and its successful processing is the real issue of life origin science.

1.2 Prescription of what?

The next question to address is, "Prescription of what?" The answer is: organization, function, utility, usefulness, computational success, and intuitive pragmatic work. None of these phenomena are physical. They are all abstract, conceptual, nonphysical and formal. They can have physical manifestations. Formalisms can be instantiated into physicality. But phenomena like "organization" remain fundamentally formal, not physical. They are choice-based. The laws of physics can spontaneously generate order—the subject of chaos theory. But physical laws cannot possibly generate formal organization or sophisticated function (See Sections 2.10 – 2.15). It is astounding how many otherwise brilliant scientists remain blind to this simple fact.

A prebiotic environment was devoid of consciousness. Prescription could not have been programmed at decision nodes by inanimate nature. No "agents" existed to follow directions, either. Information, including Prescriptive Information (PI), would have been worthless with no sentience, desire, intent or choice-contingency to obey its instructions.

Function had to be valued, sought out and preserved for metabolism to organize. This is why pursuing the topic of "information" in life-origin science is a red herring. Abiogenesis research must address the problem of *direct causation of organization and integrated function*. This is what we mean in this book when we talk about Prescription and its processing. We are talking about *direct causation*, not just "information about." The birth of both prescription and its processing is the real issue of life origin.

How did genomic and epigenomic instructions come into existence prior to any naturally selectable fittest phenotypic function, let alone living organisms? Prescription, programming and its processing arise only out of "choice determinism" rather than physicodynamic determinism.

Even if instructions somehow came into existence physicochemically, what would have followed (pursued and obeyed) those instructions? How did biochemical pathways, cycles, and any protometabolism get conceptually steered and organized?

Not all causes are physical. Sophisticated function/utility, even within the physical world, is always the effect of formal causation. Once again, by formal, we mean abstract, conceptual and nonphysical rather than physicodynamic or physicochemical. Formalisms always involve unconstrained choice-contingent controls. Mathematics, logic theory, computer programming, language, categorization, tabulation of results, and the drawing of scientific conclusions, are all examples of formal rather than physical causation. No naturalistic scientist can deny this. To deny nonphysical mathematical formalism is to deny physics itself. The physical laws are nearly all mathematical equations and inequalities. Measurements are formal representations of physicality, not physicality itself.[8-18]

The mathematical equations that we call "laws" precede and, in a certain sense, govern physicochemical interactions. Of course, the equations themselves don't really *do* anything. They are just representational and descriptive of "law." But the underlying Prescription behind those "laws" and their equations not only constrains, but *controls,* the unveiling of physicality. From this cosmogonic perspective, the laws were not "born" in the Big Bang event. The mathematical laws were *prescribed* prior to, and first realized physically in association with, the Big Bang event.

Cosmogony addresses the origin of the cosmos, along with the origin of its formal controls. Cosmology, on the other hand, merely presupposes the existence of the cosmos. Cosmology studies the history of the cosmos since existing, along with its current workings.

Why does physicality "obey" nonphysical, formal, mathematical laws? How did these formal mathematical equations come into existence? At issue is cosmogony, not cosmology.

The most fundamental principle of science is *not* a law of thermodynamics. It is The F > P Principle, The Formalism > Physicality Principle.[19] This Principle is discussed in section 3.16.

The *real* problem of abiogenesis (life origin) has to do with *how Prescription of formal integrated function, and the processing of that prescription, arose* in a chance and necessity, mass and energy only, naturalistic world.

1.3 Information itself doesn't *do* anything.

Information can be used in causation. But information is not synonymous with causation. A great deal of information can exist, both in its abstract nonphysical form, and in its instantiated physical form (e.g. in electromagnetic media of retention and transmission), with zero causation. Even when information is a factor in causation, it must be processed to contribute to that causation. The instantiated Prescriptive Information (PI)[2,20-22] must be *acted upon* for formal function to result. This processing of information renders PI an indirect contributor to the causation of formal function. Not only must the functional information itself be prescribed, but the processing concept, rules and machinery must also be prescribed.[23,24]

For the existence of prescription to be validated, the purpose and goal of prescription must come to fruition. Prescription, therefore, entails much more than just information.

The primary cause of PI (e.g., programming of computational function) flows ultimately from such nebulous factors as desire, intent, concept, goal, choices and plan. None of these entities are physical. They are all formal. They can only become physical causes when instantiated into physical implementations of physical causation. At that point, energy consumption inevitably becomes a factor. But in their abstract pre-physical form, the constraints of mass and energy, and the law-like orderliness of physical interactions, are irrelevant to prescription of formal function even in a material world. It is only blind belief that presupposes formalism to be an effect of physical causation. This notion is not scientifically supportable. It is also logically impossible. Physical bricks cannot construct a building on their own. Carbon, hydrogen, oxygen and phosphorus molecules cannot generate brain or mind on their own. Formalisms must first organize and control physical molecules for a brain or mind to come into existence. Only then can thoughts, desires, and purposeful choices come into existence.

Of special interest in this book is the prescription of sophisticated *initial* bio-functions in a prebiotic cosmos. To address the question, "How did life begin?" requires first addressing, "How did prescription and its processing begin?"

1.4 Information discussions lead to an epistemological quagmire

Why are information discussions a red herring in life-origin research?

Human epistemology (the study of human knowledge) is irrelevant to objective, historical, ontological life origin. Information is for potential learners and knowers.[25] There were no learners or knowers when life began in the cosmos.

Information is always *about* something. "*Aboutness*" is an epistemological descriptive concept, not an objective, historical, ontological cause. If there were no learners or knowers in a prebiotic environment, no epistemology existed in that environment. Ontological being did not include "aboutness" in an inanimate environment. Aboutness, uncertainty, knowledge, "surprisal," mutual entropy, searches for targets, and understanding, therefore, are all irrelevant to the emergence of objective, ontological prescription and its processing.

If all known life is cybernetic, cybernetics (control mechanisms) had to have been programmed and processed into existence.[26-28] Mere "information about" (Descriptive Information [DI],[2,20,21,29] or even "instructions for how to" (Prescriptive Information (PI), would not have existed. Even if they had, no sentient beings existed in a prebiotic environment to understand description or instructions. We are faced with the problem of prescription and its processing having to be entirely ontological (objective), not epistemological and subjective. Yet, the prebiotic physical environment itself is logically incapable of prescribing or programming any computational function. Immediately, materialism/physicalism/naturalism as a metaphysical worldview conglomerate would seem to be untenable. How could the purely philosophic axiom, "Physicality is sufficient," possibly hold true in the real world of cybernetic life? It is incumbent upon philosophic naturalists to explain and defend their metaphysical belief system, especially if they are going to try to contend so fanatically that physicalism is "proven scientific fact." How could non-sentient physicality produce the readily observable ontological *programming* of life?

The argument that "Life was much simpler then," will be deconstructed from many different angles throughout this book. The problem of Prescription and its processing applies to even the simplest Sustained Functional System (SFS)[29-32] (See Sections 2.4.4. and 3.11), including the simplest conceivable protocell model (remembering that protocells have to be alive, not just soap bubble-like entities illegitimately imagined to be objectively alive).

Current notions of "information" are confused at best. Notions of information are often so erroneous and firmly entrenched in folly, even among information theory specialists, that it becomes pointless to try to untangle the massive knot. It is like a hopelessly tangled fishing line. Just "cutting bait" is far wiser. The entire entangled informational nightmare needs to be trashed.

Information, including "Functional Information" (FI),[4,33-38] and even the more focused concept of Prescriptive Information (PI),[2,20,29,39] are a "can of worms." They both involve too heavy a dose of inseparable human epistemology. Human knowledge only beclouds the issue of what caused abiogenesis. Humans didn't exist then. Whatever happened 3.7 billion years ago (?) had no dependence upon human consciousness.

There were no other organisms, either, in a prebiotic environment, let alone multicellular organisms. No vestige of a central nervous system existed to learn from, or be instructed by, epistemological information. Shannon uncertainty, "surprisal," mutual entropy, knowledge, probability considerations, and the pursuit of targets, therefore, have nothing to do with life-origin.

The cell controls and regulates its own metabolism and survival, and had to have done so from the first primordial cell on. How do cells achieve homeostasis? Circuits of ingenious configurable switch-settings were required prior to realizable phenotypic structure and function. Physical symbol vehicles (tokens; nucleosides) had to be arranged in certain syntax according to arbitrarily selected formal rules (not laws). Three-dimensional architecture needed design and engineering for relational structure to provide formal utility (e.g. ribozyme catalysis of hundreds of crucial metabolic functions). All of these integrative functions had to be *prescribed.* Prescription of Function (PoF) requires abstract, choice-contingent concept and goal. Concept arises only from nonphysical formalisms, not from the chance and necessity, mass and energy, of physicality. PoF not only precedes, it causes the effect of formal utility. PoF generates phenomena such as computer programming, architectural design, engineering specs, even genomics.

Ontological PoF is a more focused subject of study in biology than Prescriptive Information (PI). Prescriptive Information (PI) is a cut above "Functional Information" (FI).[4,33-38] in describing genomics and epigenomics. But, PoF surpasses even PI, because it is:

1) objective rather than subjective

2) ontological rather than epistemological

3) efficacious in actually producing cybernetic cells, rather than just instructing or describing them.

If there were no physical brains or minds in a prebiotic environment, how did inanimate nature *prescribe and process* the first organized bio-function? Is there something in physical law that could organize integrative bio-function? Can random heat agitation of molecules organize bio-function? Can quantum mechanics integrate macroscopic circuits and organize a protometabolism?

Which of the four known forces of physics organized and prescribed life into existence? Was it gravity? Was it the strong or weak nuclear force? Was it the electromagnetic force? How could any combination of these natural forces or force fields program decision nodes in pursuit of eventual formal utility?

Why would a prebiotic environment value, desire or seek to generate utility?

Can chance and/or necessity program or prescribe sophisticated bio-function?

Science is a human epistemological system. It is admittedly impossible to divorce our knowledge, with all of its problems, from scientific investigation. But those problems can be greatly minimized by shifting the emphasis away from "uncertainty reduction in human minds" into the rise of primary ontological prescription in a prebiotic environment. This becomes a matter of simple direct causation rather than a quagmire of epistemological confusion.

Science has very good ways of minimizing the human epistemological problem:

- Double-blind studies
- Independent groups all performing the same experiment
- Conference debates
- Letters to the editor
- Prediction fulfillments
- Universality of application of theories, laws and paradigms
- Falsifiability
- Many other checks and balances.

Life origin, wherever in the cosmos, is the subject of the science known as Astrobiology. We are free to imagine abiogenesis (life from nonlife) happening on earth, or on some other planet or moon in any solar system. Forget all about information for the moment. Think *Causation of formal function in a physical world.* How did the tokens of nucleotides get sequenced into meaningful (functional) syntax? How did the very first genetic Turing tape get programmed? How did its Turing machine arise at the same time and place? How were the rules of translation written? How do certain cytosines in DNA get methylated so as to turn genes on and off at needed times in the future?

Thus, having "cut bait" from the bio information quagmire, we shall start anew with a fresh new fishing line. We will leave behind all of the sterile circular discussions of "mutual entropy." We will concentrate on the heart of the matter—what organized, programmed, steered, controlled and regulated initial life? What in an inanimate, prebiotic, naturalistic environment could have *prescribed integrated metabolic* bio-functions *sufficient to organize life?*

Let us step back for a moment, however, from the exceedingly difficult questions of life origin. Let us thoroughly digest what prescription is, what it requires, what it can do, and how it might have first come about. How does the processing of *any* prescription come about? Let us get a handle on the sorts of things prescription invariably requires. But let's, "Keep it simple, stupid!"

2. What does Prescription require?

We cannot address primordial prescription until we first understand the necessary and sufficient requirements of prescription. Thus, we must begin our investigation into the origin of life with an analysis of *The Prescription Principle.*[5]

We need to expound on some of the opening summary statements found above in the Introduction. Those statements need to be supported. What does Prescription invariably require?

What are the necessary and sufficient conditions for Prescription of physical function to take place?

a) Contingency - freedom from inflexible law-like determinism. Law-like determinism is constrained. Contingency is unconstrained.

b) Choice-Contingency rather than mere Chance-Contingency

c) Intent—purpose and goal

d) *Selection FOR potential* function rather than mere Selection *FROM AMONG existing* function

e) Programming proficiency—the ability to successfully pursue and achieve optimization of sometimes parallel computational paths to success. The ability to anticipate what will work and to effectualize choice sequences into optimized function.

f) Instantiation of prescriptive choices into a material medium

g) Ability to design and engineer equipment/machinery/computers needed to *process* that instantiation.

h) The ability to communicate wise programming choices linguistically. The ability to use a coded material symbol system to instruct and send meaningful and interpretable messages to a distant location.

Choice entails decision nodes. The simplest form of decision node is binary, where one must choose from among two options (Should we Open or Close a configurable switch, and for what reason?). Such a bifurcation point, or fork in the road, becomes a true decision node only when one path is purposefully chosen over the other. If we flip a coin to decide which path at the fork to take, that is not a legitimate decision node. The direction we take at that bifurcation point would then be random. No true choice is made. How efficiently or efficaciously we arrive at our desired destination using nothing but coin tosses is perfectly predictable probabilistically: "no better than chance." Randomness never programmed any sophisticated function, nor did it assist us in the slightest in arriving at a desired destination in the shortest distance or quickest time.

Note that the necessary and sufficient conditions for Prescription are all formalisms, not mere physicodynamic cause-and-effect interactions. Let us examine more details of prescription.

2.1 Prescription, unless we pioneer it, has to be instructed with prior purposeful choices.

Instructions can be generated "from scratch." They can be original instructions. But, original instructions can only arise from bona fide decision-node choices. The issuance of any form of instructions is always a function of Decision Theory, not Stochastic Theory. Guiding instructions have never been observed to arise from the chance and necessity of inanimate nature.[40] They have only arisen from Choice Determinism (CD) at true decision nodes. Decision nodes are independent of cause-and-effect Physicodynamic Determinism (PD).

Usually, instructions are pre-existing. We just follow them for efficiency sake, rather than trying to "re-invent the wheel" ourselves. But when instructions are pre-existing, the question remains, where did the pre-existing instructions come from? At some point, pre-existing instructions had to have been generated by purposeful choices, not by chance and necessity. Instructions are always formal, never physical.

2.2 Prescription includes not only Instructions (Prescriptive Information [PI]), but the *processing* of those instructions.

Not to belabor the point, but for reasons of organization and sensible outlining of this book, we have to restate, in summary form at least, some of the points already made in the introduction.

The *processing* of algorithms required to convert instruction into pragmatic physical reality also arises only out of CD, not PD. Instructions are worthless without the ability to process them. For prescription to become reality, processing is required. Instructions must be understood, and used for prescription to be realized. The desired end-product for which prescription was given must become reality. This requires processing of those instructions.

The workings of a theoretical Turing machine provide an example of just how sophisticated is the processing of a Turing tape needed to process a Turing tape of already existing computational prescription. This is ironic, since the Turing machine supposedly represents the most reduced or simplistic model of PI processing. Yet, if the workings of a Turing machine were included, or even briefly reviewed here, most readers would abandon reading this book because the subject matter was too dense.

2.3 The processing of Prescriptive Information (PI) itself requires PI.

Any organizational and/or utilitarian process involves a goal, set-up "know-how," directionality, steering, control and on-going regulation. All of these aspects of formal "process" require instruction. Processing instruction is itself a form of Prescriptive Information (PI).[2,22,31,41,42]

The "how-to" of processing involves rules. Rules are not laws. Rules are arbitrarily chosen rather than forced. Rules can be disobeyed at will. But disobeying "tried and true" rules usually results in compromised algorithmic function. Formal design and engineering are also required for physical machines to be able to process non-trivial PI, especially optimally.

Turing machines don't just spring into existence spontaneously, any more than Turing tapes. Such a conceptual device has to read and execute the string of instructions. For a Turing machine to come into existence, it must be designed and engineered to perform several different kinds of minimally desired tasks. These supposedly simple desired tasks are much too sophisticated for any spontaneously-occurring simple machine to achieve. An inclined plane, for example, is one of the "simple machines." A hill can spontaneously form naturalistically. But a hill does not become the simple machine of an inclined plane until an agent *uses* that hill to accomplish *useful* work. Just because a tumble weed is pushed up the hill by the wind does not mean that hill is a machine. If this is true of a simple inclined plane machine, how much more evident is it of a Turing machine? The Turing machine must be designed and engineered by an agent to read and execute a Turing tape. But even more significant, a Turing machine is not just a simple machine. To perform its desired function requires a great deal of goal-oriented Prescription and Prescription Processing. Reading and executing instructions is itself formal rather than a cause-and-effect law of nature. And, certainly, the creation of the Turing machine that accomplishes this task is formal, even though it uses physical components.

2.4 Prescription comes in many forms.

2.4.1 Instructions/directions using semantic language expressed by using a formal Symbol System.

Unless we are pioneering our own new culinary delights, we typically request a recipe that tells us how to prepare a tasty dish. A recipe is a recordation of prior purposeful choices. That recipe consists of a list of needed ingredients, and linguistic instructions of how to combine and process those ingredients to produce a desired product.

2.4.1.1. Non-material symbol systems

Prescription and programming are typically recorded, and usually encoded, into a formal representational symbol system. Each symbol represents a different discrete choice from among at least two possible options. For example, either a

- "0" or a "1," with no decimal places in between.
- "Yes" or a "No," with no gray area in between.
- "Open" or a "Closed" binary configurable switch, with no half-on or half-off.

This kind of symbol system provides a logical "excluded middle." It provides a discrete free decision that is unforced by physical determinism.

Decision nodes can also have more than two options to pick from. A quaternary decision node, for example, requires a purposeful choice of one discrete option from among four options. Such a decision node would correspond to two bits of Shannon Uncertainty, prior to selection. Once one of the four options is actually chosen, no Shannon Uncertainty (zero bits) remains. This is important for investigators to understand who are still confusing Shannon Uncertainty with bona fide intuitive, semantic, Functional Information (FI).[2,21,22,31,42] A completed choice contains 0 Shannon bits of

Uncertainty. Contrary to common belief, binary choices themselves cannot be measured with bits! Only the number of binary choice *opportunities* can be measured with bits. When bits are used to supposedly measure already completed programming choices, they are in reality only measuring the number of place-holders that were required to record those choices by the program, not the specific choices themselves.

Most symbol systems provide a large alphabet of symbols from which to pick. Language is a prime example. Each symbol represents a formal, nonphysical choice from among real symbol options. In language, each letter is picked from an alphabet of letter symbols. English, for example, would have 26 options at each decision node, excluding punctuation options.

It is worthy of repeating that the Selection *FROM AMONG* each symbol always represents *a purposeful choice.* Using a mere coin flip to decide what binary symbol to use is not a programming choice. Mere Shannon combinatorial Uncertainty and random selections cannot provide instruction for sophisticated function. Mutual entropy (reduced uncertainty) and "ambiguity" cannot prescribe either.

Symbol systems confined to one's own mind remain formal, the same as "doing math in one's head." The choices and instructions themselves are abstract and conceptual. A prevailing metaphysical presupposition is that mental choices are ultimately produced by physicodynamic cause-and-effect (by the physicality of brain) alone. We will be carefully examining this pre-assumption from many different perspectives throughout Section 3 of this book.

2.4.1.2 Material Symbol Systems (MSS)

To communicate formalisms to others, those formalisms have to be represented and instantiated into some physical format. To accomplish this, programming choices, and their representative symbols, are most often recorded into physical reality using *physical* symbol vehicles (tokens) in

Material Symbol Systems (MSS).[43-45] Choice Determinism (CD) is introduced into physical reality through the choice of each physical symbol vehicle. This converts formal representationalism into physical reality. The purposefully chosen tokens, and their sequencing (or syntax), according to formal rules, allow prescription to enter the physical world for memory retention and transmission. Fellow Scrabble players across the game board will be critiquing your token syntax for conformance to the rules of Scrabble.

In addition, this physical representation (e.g., photons sent via fiber optics) has to be transmitted through a Shannon channel to a receiver and destination. The only way the recipient can read and understand the message, and get any functional benefit from that message, is for both sender and recipient to use the same set of arbitrarily-chosen formal rules to write and interpret the meaning of the symbolic message. The message is now instantiated into physical photon sequencing, a Material Symbol System (MSS).

Remember, since mass and energy are interchangeable ($e = mc^2$), whenever we are addressing energy needs and consumption, we are talking about physicality, not formalism. Formalisms are nonphysical. Formalisms have no dependence upon mass or energy. Consider the existence of mathematics (e.g. pi) and the birth of the numerical force constants and mathematical laws of motion immediately prior to the Big Bang. We cannot attribute mathematics to a physicodynamic reality that had not yet been born.[19] (See Section 3.16 on the most fundamental principle of science, The F > P Principle).

Sequences of choices can be represented by sequences of symbols, otherwise known as *symbol syntax*. Symbol syntax makes Prescriptive Information (PI) potential seem limitless. It is impossible to dichotomize syntax from message meaning when it comes to discussing message pragmatics (utility) and apobetics (purpose).[46-48]

The physical nature of these tokens (physical symbol vehicles) can vary considerably. Tokens may take the form of ink molecules on paper (e.g., a handwritten, typed, or printed string of molecular-conglomerate tokens). They can be physical tokens in the form of electromagnetic regions on a hard disc. They can be photons or electrons that are faxed, uploaded / downloaded or emailed. The tokens can be puffs of smoke which make up smoke signals. They can be lettered blocks of wood in a Scrabble game. The tokens themselves are physical. But the choices of those tokens, and their sequencing (syntax), are nonphysical preferences of mind. It seems silly to have to keep repeating something so obvious. Yet the naturalistic scientific community regularly manifests utter oblivion to, or obstinacy against acknowledging, these simple facts. Programming selections from among multiple real alternatives are not, and cannot be, militated by physical laws.[17,36,49-58] Programming is impossible without purposefully choosing from among real options at bona fide decision nodes.

Ultimately the message is not in the physical symbol vehicles (tokens) themselves, but in their syntactical arrangements according to formal, arbitrarily-*chosen* rules. Thus, the information on this page cannot be reduced to only the physical molecules of ink and paper. Physicality is used. But, the message and prescription are not physical. The tokens are physical, but the message arises only out of purposeful choices of those tokens from an alphabet at true decision nodes. Meaning is determined by how the chosen symbols are sequenced, or arranged in three-dimensional prescriptive systems.

2.4.2 Programming (e.g., the theoretical Turing tape)

A program is a sequence of previously-made, formal, purposeful choices. When we buy an operating system or software, these choices successfully prescribe automation of particular tasks. Of course, those programming choices must still be processed.

A program solves certain problems. First, a problem must be formulated. The formulation of a problem is itself a non-physical formalism

(See Section 2.12). Conceiving a computational solution to the problem is also formal, not physical. The choices must be read and processed to realize sophisticated utility. The processing itself must also be formally prescribed.[23,24] Optimization of that solution is again formal.

So-called "evolutionary algorithms" and "directed evolution models" are riddled with purposeful steering, goal-orientation, and other non-naturalistic formal components. They only achieve usefulness through agent-controlled experimental design and investigator interference in a supposedly naturalistic process.

Notice that the choices at each decision node must be made in advance of computational success (in advance of any function). No environmentally selectable usefulness exists at the time each programming choice must be made. Yet the final computational success depends squarely on those choices. Programming predetermines computational success, the "halting problem"[1] notwithstanding. We may not be able to prove that a certain program will finish its task and "halt" without actually running it. But that doesn't change the fact that the success or failure of the computational program depends upon a lot of prior decision-node purposeful choice commitments. This is the essence of programming, and the required choices are formal, not physicodynamic (physical: limited to mass/energy interactions, force fields, the laws of motion, starting conditions, and time). Formalisms are independent of all of these physical constraints.

To implement an algorithm into physical reality requires:

1) Use of a material symbol system (MSS)[43,44,59] (See Section 2.4.1.2),

2) The integration of configurable switch-settings into a circuit, or

3) The engineering of component parts and their assembly into an organized Sustained Functional System (SFS)[29,30] (e.g., a machine) (See Section 2.4.4 and 3.11).

None of these functions can be performed by physicochemical reactions or phase changes. They are formally implemented, without exception. The recorded choices must be recorded, or instantiated, into physicality.

Both computers and cells use a combination of all three of the above means of instantiation of Prescription into physical utility.

As stated above, programming choices are not accidental commitments. They are not random "coin flips" at mere "bifurcation points." And they are not militated by physical law, either. They are true choices at bona fide decision nodes. Any attempt to disallow purposeful choices in the explanation of non-trivial programming results in a "blue screen" every time!

Programming and its processing is an ingenious engineering discipline that seeks to prescribe an automated optimal computational path before it exists. Programming is more like an art form than the physicodynamic interactions of inanimate nature. No one has ever observed the laws of physics generate a computational program. Formalisms may all wind up being instantiated into physical media for memory, transmission, processing, and physical utility. But they are all fundamentally nonphysical, abstract and conceptual.

Programming can also provide a form of "how-to" instruction that provides to end-users sequential discrete pre-made, previously-proven-successful decisions. Whatever specific form it takes, programming always consists of making *purposeful* choices. There can and will be no exceptions to this rule. It is a logical necessity, a deductive absolute, that non-trivial programming arises only out of free-will choice-contingency.

An algorithm must be coded using a target programming language. Formal logic is required. A source code must be written and "implemented" in a way that "makes sense" to the processor. At the most fundamental level, this source code is usually written in binary code. A choice must be made between a 0 and a 1 at each binary decision node. Most computer systems use groupings of eight binary choices called a "byte." More complex code than that is usually built using hexadecimal (16) and higher groupings (e.g., 32).

It is hard for humans to write code using long strings of 0's and 1's. So, a set of mnemonics in an assembly language is used to direct computer processors. This approach allows programming instructions to make more sense to human programmers.

Higher level programming uses gates and logical operators, such as "and" operations, to convert the code into more common-sense language and commands. Everything hinges on wise, purposeful choices.

Software must have "hardware" (or possibly "firmware," or even "wetware") capable of reading and processing the instantiated PI on a Turing tape. A theoretical Turing tape provides means of physical recordation, storage, and access of instructions for processing. Some of the data on that tape instructs the Turing machine operation itself. Other data provides valuable PI for the Turing machine to act on. Sometimes a Turing machine can even change the data on the tape. The distinction between hardware and software has become progressively blurred in modern computing. Both hardware and software are choice-determined, not physically determined. (See Section 2.4.2, 3.21 and 3.22.)

2.4.3 Pre-set configurable switch-settings in an integrated circuit (e.g., chips, circuit boards).

Another way to represent and record formal purposeful choices is through the prior setting of configurable switches. The configurable switches themselves are physical. But the setting of these switches is entirely formal.

Yes, it takes a physical force to push a switch knob in one direction or another. The same is true of Maxwell's demon's trap door (See section 2.14, 3.3.2, 3.11.4). But, the *choice of direction* is entirely formal, not physical.

Whichever way the switch knob is pushed requires the exact same amount of energy. The direction of switch-setting is physicodynamically indeterminate, neutral and inert.[45,60] It is Choice Determined (CD), not Physically Determined by the laws of physics and chemistry (PD). The laws of physics are indifferent to which direction the switch knob is pushed. This spells "freedom" from physicodynamics. This freedom is the essence of what makes a configurable switch configurable, and why configurable switches are so important in the instantiation of nonphysical formalisms into physicality.

Figure 1 this Section, 2.4.3, shows the use of configurable switch-settings and an integrated circuit to instantiate nonphysical, formal choices.

Figure 1. A row of dip-switch settings depicts a form of prescription using sequential configurable-switch settings. Syntactical binary choices can control programmable integrated circuits. Choice Determinism (CD) is incorporated into purposeful physical configurable switch-settings that collectively prescribe and integrate formal function. (Used with permission from: Abel DL: The capabilities of chaos and complexity. *Int J Mol Sci* 2009, 10:247-291.)

Thus, the setting of a configurable switch is not only a way of *representing* a formal purposeful choice, but it is also a way of prescribing that choice into the physical world. The integration of switch-settings makes integrated circuits possible. This in turn alone makes circuit boards, computers and electronic devices possible.

No laws of physics are violated in the programming of configurable switches. Yet the effects of the particular functional settings of these configurable switches cannot be reduced to laws and constraints. Their functionality stems directly from their formally chosen settings. This constitutes the only known mechanism of bona fide controls. Configurable switches are the key to escaping the bounds of low-informational (highly constrained and ordered) physicodynamics to soar into unlimited formal creativity. Programmatically set configurable switches are also the key to exceeding the relative pragmatic uselessness of chance-contingency.

2.4.4 The integration/organization of component parts into a holistic Sustained Functional System (SFS).

A Sustained Functional System (SFS)[29,30] (See also Section 2.4.4 and 3.11) is any conglomerate of entities that collectively perform some non-trivial function for an extended period of time.

Organization always requires purposeful choices. Organization must never be confused with spontaneous self-ordering in nature (crystallization, chaos theory self-ordering such as tornado and hurricane formation). No "dissipative structure" in chaos theory requires a single purposeful choice to generate. Organization's very essence is the practice of choice with the intent to achieve some pragmatic goal. Inanimate nature is blind to formalisms such as "usefulness." The environment has no interest whatever in "function" or "utility."

The most common realization of Sustained Functional Systems (SFS) is what we would call "machines." The integration/organization of component parts into a holistic machine is a representation and manifestation of Prescription. No non-trivial machine has ever been observed to spontaneously generate, apart from choice-contingency at bona fide decision nodes. A machine is a sort of three-dimensional prescription, rather than mere linear digital programming.

We frequently misuse the word "system" to refer to things like weather fronts. But, weather fronts are not true systems. Systems are formally organized by choice-contingency to accomplish some pragmatic task. A weather front is an interface of different temperatures, pressures and phase changes. But a weather front involves no formal choice-contingency, and achieves no non-trivial utility. It is entirely inappropriate to refer to a self-ordered physical interface as a "system." It has no formal components, except that the laws of physics themselves are formal, owing to the F > P Principle discussed in Section 3.17.

2.4.5 The design/engineering/manufacturing of those individual component parts needed to make any non-trivial machine or other Sustained Functional System (SFS).

Included in SFSs[22,29-31] are architectural arrangements of physical structural components that achieve formal utility. Technically, a building is a system. It came into existence only through purposeful organizational choices of component parts. Not only does the building manifest non-trivial utility, but each component part does also. The component parts must also be designed and engineered. The component parts don't just spring spontaneously into existence. Finally, a building doesn't build itself. Buildings only come into existence by Choice Determinism (CD).

Man-made machines are also examples of SFSs. But man-made machines are not the only machines in the world. Many thousands of subcellular molecular machines are, alone, what make life possible. Section 3 of this book addresses the origin of molecular machines, along with the organization of cells and the prescription of their many processes.

2.4.6 Pre-programmed degrees of end-user freedom of choice.

Computers and software are frequently designed to grant end-users considerable degrees of freedom to choose which operations the user wants to pursue. This end-user freedom to control computational paths is itself a representation of prescriptive prowess. It adds greatly to the sophistication of programming for users to be able to choose which path to functionality is of interest.

The point to remember is that even end-user freedom opportunity must be prescribed by programmer choices.

2.5 Prescription and its processing invariably pursue the *goal* of function.

Prescription requires *Selection FOR* (in pursuit of). Programming is pursued only when we have some desire or need for utility. A pragmatic goal to obtain that utility results. Computational success, for example, facilitates reaching that goal.

Evolution offers only Selection *FROM AMONG* (existing already-programmed, already-living phenotypic organisms). Evolution has no goal.

Selection *FOR* always involves pursuit of a goal. Both Selection *FOR* and the pursuit of a goal are always abstract, conceptual and formal, not physicodynamic.

2.6 Prescription can only be verified through successful realization of utility.

For the existence of prescription to be validated, the purpose and goal of prescription must come to fruition. Prescription, therefore, entails much more than just information. The *effect* of function that has been *caused* by prescription must be empirically verifiable.

Even Prescriptive Information (PI)[2,20] requires *processing* before its instructions can produce utility.[23] (See Section 2.2 and 2.3). Computational or manufacturing success is required. Just talking about information fails to address this fact. Prescription involves the whole package of *generating* sophisticated utility. A world consisting only of chance and unimaginative necessity (e.g., the laws of motion) cannot prescribe and process sophisticated utility into existence. Mere mass/energy conversions and chemical reactions possess no ability to prescribe formal organization or non-trivial functionality.

2.7 Prescription requires contingency. But, there are two kinds of contingency.

Before we can expound on prescriptive causation, we must first review what scientific literature has to say about **contingency**.[6,7,19,21,22,29-31,42,61,62] Contingency means that events can occur in multiple ways despite monotonous physical law, initial condition constraints, and probability bounds.

The steering of prescription toward non-trivial functional success requires *choice-contingency,* not just chance-contingency. *Purposeful choices* must be made in pursuit of non-trivial utility. Thus, we must address the two different kinds of contingency: 1) chance-contingency and 2) choice-contingency.

2.7.1 Chance-Contingency

Chance-contingency is unchosen and undirected toward any goal. No preference of direction or steering exists toward pragmatic benefit with chance-contingency. Only the most naïve "function" arises by chance. Usually, even that naïve function can only be called "function" if some agent decides to make use of it.

Chance-contingency is what we seem to observe in statistically describable quantum events and in the molecular collisions of heat agitation. In the latter, uncertainty is high as to what will happen despite known causal chains.

Sproul gives an excellent account of the inability of chance to cause any physical effect.[63]

Most theorists attempt to reduce chance-contingency to unknown and/or complex causation as summarized by Peale.[64] Thus chance-contingency may be only "apparent." In any case, no deliberate selection from among options occurs with chance-contingency.

2.7.2 Choice-Contingency

The steering of events towards non-trivial functional success requires *choice-contingency,* not just chance-contingency. *Purposeful choices* must be made in pursuit of non-trivial utility. Prescription is impossible without choice-contingency. The physical interactions that are militated by cause-and-effect law are not programmable. Programming requires freedom from law. The configurable switch-settings needed to integrate circuits, for example, must be freely selectable. If the laws of physics and chemistry forced those switch-settings into the same position every time, by law, programming creativity and ingenious computational function would become impossible. No programming would be needed if ingenious function happened by law.

Programming and Prescription are invariably a function of Choice-Contingent Causation and Control (CCCC).[5] Philosophic naturalism makes myriad attempts to tap-dance around this undeniable aspect of reality. Metaphysical naturalism will do back flips in a futile effort to recast CCCC into chance and necessity. The worldview of naturalism lives in fear of having to acknowledge what has also been termed Choice Determinism (CD). CD is distinguished from law-like physicodynamic and physicochemical determinism in two major ways: First, CD is never automatic or spontaneous in inanimate nature. CD does not and cannot arise from mere chance and/or necessity (natural law).[65] Second, CD can only arise from purposeful choices that steer or direct behavioral outcomes toward the goal of some desired function.[66]

CD is not always pragmatically wise. Bad purposeful choices can be made, with predictable results! But CD is normally exercised under the desire and belief that purposeful choices will yield more beneficial results than random events, as valued by some agent.

Only "agents" are known to value anything. Only agents are known to pursue attainment of such value. Agents alone pursue functionality and usefulness, not inanimate environments.

Prescription of Function (PoF) exists in the abstract prior to its instantiation into physicality. Such instantiation is often two-step. First, the instructions must be recorded and stored in a physical medium for easy access and processing by physical machinery. Second, those instantiated instructions must be processed into physically realized functionality. Component parts must first be manufactured according to recorded instructions. Next, they must be conceptually organized and assembled in a certain functionally-integrated way into physical processes. Formal computation must be performed by formally integrated circuits and configurable switch-settings that serve as true *logic* gates. Logic theory is formal, not physical, even though the configurable switches of the processors are physical.

Open vs. Closed must be purposefully *chosen*,[41,61] not just statistically described as a measure of combinatorial uncertainty.[67]

Physically instantiated instructions employ electromagnetic flux in computers. Energy expenditure is required in a physical world to realize and experience the benefits of computation. Once again, formal prescription of function precedes instantiation into physicality. Formal prescription is nonphysical, and therefore immune to unique energy requirements at each binary decision node of the most basic source code.

2.8 The Prescription Principle: Prescription requires Choice-Contingency.

All of the above points go into formulating *The Prescription Principle*. *The Prescription Principle* states that true organization (not just self-ordering a la Chaos Theory) of physical objects, and the achievement of non-trivial functionality, can only be achieved through the formal instantiation into physicality of wise nonphysical *Choice Determinism*.[30] Choice-Contingent Causation and Control (CCCC).[21,22,30,68] is just another descriptor found in peer-reviewed scientific literature for Choice Determinism (CD)[5] as opposed to physicodynamic (physical) causation.

As we have already discussed, for nonphysical formalisms to be able to cause physical effects, physical functionality included, those nonphysical formalisms must first be instantiated into physicality. Why do we keep covering this same point? Because the world's naturalistic scientific community deliberately obfuscates it rather than admit it. At stake is philosophic naturalism's most cherished metaphysical axiom, "Physicality is sufficient." Physicality is plainly not sufficient. The entire worldview is bankrupt because it fails miserably to correspond to *ontological being*, to the objective reality in which we all have to live.

Just because function has beneficial physical effects does not mean that function is itself physical. "Utility" is an idea—an abstract concept.

Utility is a goal invariably pursued to meet a desire or need. Inanimate nature does not have desires, goals, or ideas. Inanimate environments do not pursue "utility." A prebiotic environment could have cared less whether anything functioned. No impetus existed to progressively climb mountain peaks of optimized function from foothills of incidental molecular interactions. Reaching the mountain peak has to be valued and pursued.

Not only is a great deal of highly refined energy required to climb mountain peaks, but motivation is required. Sophisticated mechanisms are required to harness, transduce, store, and utilize free energy when needed. Cells must get usable energy without suffering oxidation or heat damage that would kill them.

We observe innumerable effects every day of our lives that could not possibly have been caused merely by the laws of physics and chemistry. Look at your automobile, your television, your cell phone. Is anyone prepared to seriously try to argue that chance and necessity, mass and energy, alone brought these phenomena into existence? Just like computers, we know intuitively and empirically that they had to have been formally prescribed and processed into existence. As a sort of corollary, we also know from universal experience that a great deal of manufacturing of parts and machines would have also been required to enable prescriptive processing.

All known living cells are programmed. They are cybernetic. Their homeostatic metabolism is highly controlled and tightly regulated at every turn. To have any hope of explaining life origin, we must first explain the birth of prescription and its processing. Then, we must explain the instantiation of prescription and its processing into the physical cosmos.

Bona fide organization cannot be pursued without purposeful choice-contingency. Self-ordering can occur without reference to organization or function; but not self-organization.[69] The latter is a nonsense term, anyway. "Self-organization" is self-contradictory. Nothing can self-organize itself into existence from nonexistence. An effect (organization) cannot be its own cause.

Choice-contingency, and the mind associated with making such purposeful choices, cannot possibly be an epiphenomenon of physicality. Chance and necessity cannot generate the third fundamental category of reality.[19] Selection from among real options, especially in pursuit of utility, cannot arise by law or chance. Law would select the same way every time. If computer programs were generated by law, they would consist of all 1's, or of all 0's, by law! Nothing would compute with a program of all 1's. No freedom of selection would exist. Programming of computational function would be impossible. Random programming produces nothing but a giant software "bug." O's and 1's chosen by coin tosses will result in a program that "blue screens" every time. Neither chance nor necessity, or any combination of the two, can replace purposeful choice.

After many decades of concentrated neurophysiological research, The Mind/Body problem remains alive and well. The reason is that "mind" cannot be isolated from Choice-Contingent Causation and Control (CCCC). CCCC cannot be generated or explained by mass/energy, chance or necessity interactions.

Similarly, the integrative utility produced by Sustained Functional Systems (SFS)[30-32] (See Sections 2.4.4. and 3.11) can arise only from wise Choice Determinism (CD), not from mere physicodynamic interactions. Chance and/or Necessity cannot generate computational programming success or sophisticated utility.[21,22,31] Nontrivial function must be prescribed by wise choice-contingency at bona fide decision nodes.[41,42,68,70,71]

"Decision nodes" cannot be reduced to mere "bifurcation points" (forks in the road) at which a mere coin flip determines which fork is taken. No evidence exists of coin flips improving the efficiency of reaching one's destination. Only wise choices at bona fide "decision nodes" get us there quicker. Rats improve their exit time from a maze only by memorizing efficacious choices from among real options at true decision nodes. These choices must be made in pursuit of the eventual, *future* function of escaping the maze. No existing function exists at the time that each decision-node choice must be made. Selection is *FOR* potential function, not *FROM AMONG* existing function. Natural Selection is always *FROM AMONG*, existing function. Natural selection is never *FOR*, in pursuit of, future function. Evolution has no goal.

Random mutations are another example of chance-contingency rather than choice-contingency. Of course, not all mutations are random. But partial constraints on mutations still do not constitute choice-contingency. Nonrandom mutations (that are not preprogrammed in the genome) model "necessity" rather than chance as their source. Necessity only serves to reduce uncertainty, which reduces PI instantiation *potential*. Neither chance nor necessity can program. Genomes consist of programmed instructions and controls.

2.9 The first step towards achieving utility is usually "organization."

What is organization?

1) Categorization: grouping by characteristics and function

2) Integration of component parts into a holistic configuration

3) The joint pursuit and achievement of utility by members of a group

4) Control and regulation of formal function

5) Tidiness

Not all of these concepts are relevant to abiogenesis, of course.

For good reason we call living organisms "organisms." They are first and foremost formally organized holistic entities of associated parts. A Unity and a coherence exist between the components. The operation of the components is integrated so as to pursue and achieve desired overall utility. Organizations always have a purpose and goal. The goal of organisms is to remain alive, to optimize their control over environmental challenges, and to reproduce. But all of these quests are pre-programmed. Certainly, mutations have no motivations. And, mutations have no ability to organize new structures or prescribe new programming.

2.10 Organization requires choice-contingency at decision nodes.

Organization is achieved only through formal Choice-Contingent Causation and Control (CCCC).[21,22,29,42] otherwise known as Choice Determinism (CD).[29] Necessity (fixed, inflexible, physicodynamic determinism) is blind to utility.

Organization must be prescribed. It is formal, not physical. An agent must choose to pursue the goal of formal utility. Self-ordering phenomena never do this. (See Section 2.4.4 and 3.11.)

2.11 Prescription is a function of Decision Theory, not Stochastic Theory.

"Decision nodes" cannot be reduced to mere "bifurcation points" if sophisticated function is expected. Prescription, and its efficaciousness, *cannot be evaluated or measured by stochastic theory.* Any form of direct causation increases probability to a theoretical 1.0. Within cybernetic systems, causation is regarded as formally absolute. The standard deviation and bell curves needed for physicodynamic causation are not functionally relevant, except with regard to the physicality into which formalisms must be instantiated in order to control that physicality.

Most naturalistic scientists believe that any phenomenon can be cast into a probabilistic model. Laws are seen to involve random variables assuming a value with a probability approaching 1.0. But Choice-Contingent Causation and Control (CCCC) (Choice Determinism [CD]) cannot be incorporated into probabilistic reasoning. CD is a deliberate, purposeful cause that lies in a different fundamental category of reality from chance and necessity. To try to apply statistical analysis to CCCC and CD constitutes a logical fallacy known as a "category error."

2.12 Formalism vs. physicality. What exactly is "formalism?"

As we have already alluded to many times, two fundamentally different categories of phenomena exist.

1) Physical: Those that can be explained by purely physicodynamic explanation (Physicodynamic Determinism [PD])

vs.

2) Formal: Formal phenomena are abstract, conceptual, non-physical, choice- and goal-contingent. Formal undertakings are caused by Choice Determinism [CD], not Physicodynamic Determinism [PD]).

All causation can be divided into these two fundamentally different categories. This distinction constitutes **The Universal Determinism Dichotomy (UDD)**.

2.12.1 Physicodynamic Determinism [PD]

Many decades ago, virtually every graduate science student and bench scientist had an immediate and clear distinction in their minds between "formalism" and "dynamics." Everyone knew that "dynamics" referred to physical interactions, chemical reactions, phase changes, physical force fields, and everything governed by the laws of motion. Disciplines like mathematics, logic theory and language were readily acknowledged to be "formalisms." Because of the illegitimate incorporation of metaphysical materialism into the very definition of science, that clear dichotomy has been largely lost.

The classic cause-and-effect chains involving initial conditions and the effects of force fields and the laws of motion are aspects of Physicodynamic Determinism (PD).

Chance and necessity, mass and energy can constrain. But they cannot control or steer toward desired functionality. Nature is blind to function.

2.12.2 Formalism and its inherent Choice Determinism [CD]

Formal means choice-contingent, or purposefully choice-determined. Formalisms are not dependent upon the laws of physics and chemistry.[19] But, they do not violate those laws, either. Formal phenomena are non-physical and choice-contingent. Formalism encompasses many forms of abstract, conceptual, nonmaterial, choice-based, deterministic thought. Decision nodes require freedom from, not dependence on, physicodynamic determinism. This freedom is sometimes called "arbitrariness." But arbitrariness is not synonymous with "random." *It simply means free from constraint, not free from controls.* A better term than "arbitrary" is "arbitrarily chosen and controlled." Choice-Contingent Causation and Control (CCCC) requires Selecting *FOR* (in pursuit of) potential function and utility. Only the exercise of choice-contingency, not just chance-contingency, at bona fide decision nodes, can provide CCCC. CCCC can also be referred to as Choice Determinism (CD).

CD can have pragmatically efficacious and efficient ramifications, or, CD can have unproductive, inefficient, or even deleterious results. Choice-contingency must not only be real, it must be programmatically wise.

Prior recorded CD and collective social experience tend to get generalized into rules suggesting the most expedient decision path to govern behavioral choices. Rules outline what choices have been proven in the past to be pragmatically beneficial guidelines for voluntary behavior. Unlike the laws of physics and chemistry, rules can be voluntarily broken, usually with pragmatically unfavorable results. We are tempted to think of rules in terms of constraints on our choices. But this confuses us. We lose track of the fundamental distinction between constraints and controls. Rules are a form of arbitrarily-chosen controls, not inexorable constraint or law. We can choose not to obey rules.

Some CD-based formalisms entail less "arbitrariness" than others. In the late 20th century, it was common for metaphysical materialists to try to reduce thought and mind to mere epiphenomena of physical interactions. This is impossible because it would be a *logical* impossibility for Chance and/or Necessity to generate Choice-Contingent Causation and Control (CCCC). Choice-contingency and Choice Determinism (CD) require freedom from pragmatically blind physicodynamic determinism. Inflexible law forces conformity and increases the probability of events unfolding a certain way. This decreases the number of bits of combinatorial uncertainty. It makes instantiation of choice-contingency into any physical medium progressively more difficult. Necessity is poison to prescription and programming. Law would force all configurable switches to be set the same way. Programming, computation and the pursuit of pragmatic benefit would be impossible with an integrated circuit consisting of all switches set to open, or all switches set to closed, by law. The integrator of any circuit must have freedom from law at each configurable switch-setting opportunity. Every configurable switch-setting, or every symbol selection in a symbol system, must be a truly free decision node.

Deny the reality of decision nodes, and Maxwell's demon dies (See Section 3.11.4). The trap door freezes in one position. The compartments equalize. Entropy maximizes. Subcellular life dies with the demon.

Formalisms include:

- Language
- Inferential, abductive and deductive logic theory
- Rules of behavior
- Mathematics
- Subjective perception
- The sign/symbol/token systems of semiosis
- Decision theory
- Cybernetics (including computer science)
- Computation
- Integrated circuits
- Bona fide organization (as opposed to mere self-ordering in chaos theory)
- Semantics (meaning)
- Valuations
- Pursuits of goals
- Pragmatic procedures and processes
- Ethics, including scientific ethics, in reporting results
- Art, literature, theatre, music, aesthetics
- The scientific method itself
- The personhood of scientists themselves

All of the above formalisms depend upon *purposeful choice-contingency* at bona fide decision nodes, rather than chance-contingency or necessity. Formalism also entails choices made *in pursuit of potential function.* Arbitrary rules, not laws, are also created and followed. The

Prescription Principle (Section 2.8) traverses all disciplines of scientific endeavor and will be found to be fully applicable to all future fields of study. Disbelievers in the Prescription Principle must be confronted and boldly challenged to falsify it. Only dogmatic blind belief arising out of the deepest Kuhnian paradigm rut in the history of science attempts to deny it. Materialistic cause-and-effect chains simply do not pursue or generate programmed computational success or sophisticated Sustained Functional Systems (SFS).[22,29-31]

2.13 Prescription always emanates from the far formal side of the great divide known as The Cybernetic Cut.

The Cybernetic Cut[41,42] is a great divide that dichotomizes all empirical phenomena into two subsets of determinism: Physicodynamic causation (physical causes operating on initial condition constraints) and Formal causation (abstract, conceptual, non-physical, purposeful, choice-based decisions). The law-like orderliness of nature along with the seeming chance-contingency of heat agitation and stochastic quantum reality lie on one side of the great divide. On the other side of this ravine lies the ability to choose, with intent, what aspects of being will be preferred, pursued, selected, rearranged, integrated, organized, preserved, and used.

The Cybernetic Cut explains how and where *formal controls,* as opposed to mere constraints, arise and penetrate the physical sphere to seize governance of physicodynamics. Traversing the Cybernetic Cut affords engineering-like ability to organize abstract concepts and to instantiate those concepts into physical reality. The far side of the Cybernetic Cut is both instructive and creative. It is controlling and managerial. The Cybernetic Cut must be crossed to program computational halting into any form of physical hardware.[41]

Table 1 This Section, 2.13, shows the difference between inanimate physicality on the one side of The Cybernetic Cut, and those aspects of reality originating from the opposite, formal side of The Cybernetic Cut from which all sophisticated function and pragmatic controls arise.

Physicodynamics	Traversing the Cybernetic Cut
Physical	Nonphysical & Formal
Incapable of making decisions	Decision-node based
Constraint based	Control based
Constraints just "happen"	Constraints are deliberately chosen
Natural-process based	Formal prescription based
Forced by laws & Brownian movement	Writes and voluntarily uses formal rules
Incapable of learning	Learns and instructs
Product of cause-and-effect chain	Programmer produced
Determined by inflexible law	Directed by choice with intent
Blind to practical function	Makes functional things happen
Self-ordering physicodynamics	Formally organizational
Chance and necessity	Optimization of genetic algorithms
No autonomy	Autonomy
Inanimacy cannot program algorithms	Programs configurable switches
Oblivious to prescriptive information	Writes prescriptive information
Blind to efficiency	Managerially efficient
Non creative	Creative
Values and pursues nothing	Values and pursues utility

Table 1. Comparison of the two sides of the great ravine dividing reality known as The Cybernetic Cut. All phenomena owing their existence to inanimate physicodynamics lie on the near side of the ravine. Those aspects of reality that can arise only from the far formal side of Choice Determinism, sophisticated function, and pragmatic control are contrasted.
(Used with permission from Abel DL. The 'Cybernetic Cut': Progressing from Description to Prescription in Systems Theory. *The Open Cybernetics and Systemics Journal.* 2008; 2:252-262 Open Access)

2.14 Prescription always flows across The Cybernetic Cut on the one-way Configurable Switch (CS) Bridge from formalism to physicality.

The Configurable Switch (CS) Bridge is a one-way bridge over the Cybernetic Cut.

The formalisms that can arise only from the far side of The Cybernetic Cut must traverse The CS Bridge (Configurable Switch Bridge) from the far formal side to the near physicodynamic side of this great ravine.[42] The CS Bridge is the only known means of traversing The Cybernetic Cut. Traffic flows only from the far side of formalism generation into the near side of physicodynamic mass/energy interactions. The physicodynamic near side of the Cybernetic Cut is never observed influencing, in any way the far side category of reality known as formalisms.

The setting of the configurable switches and logic gates with Choice Determinism (CD) allows formal choice-contingency to introduce formal causation into a physical world. The setting of these configurable switches constitutes the building of the Configurable Switch (CS) Bridge[42] across the vast ravine of materialistically untraversable Cybernetic Cut.

Physicality (chance and necessity; mass and energy) never prescribes formal function. The chance and necessity of physicality cannot steer objects and events toward formal utility. Chance and necessity cannot compute or make programming choices. Mere constraints cannot control or regulate. The inanimate environment does not desire or pursue function over non-function.

Conversely, non-physical formalism itself can never be physical. Mass and energy expenditure are irrelevant to any formalism. Use of the one-way CS Bridge makes possible the instantiation of nonphysical, formal Choice Determinism into physical reality. Specially designed and engineered configurable switches can only be set to open or closed positions by choice-contingency, not by chance and/or law-like necessity. The choice of physical tokens from an alphabet of tokens in a material symbol system is also means of instantiation of nonphysical choice into physicality. The manufacture and integration of component parts into a holistic functional structure is still another means of crossing the one-way CS Bridge.

How does physicality ever get organized into usefulness of any kind? How does stone and mortar ever become a building? The answer lies in our ability to build a CS Bridge from the far side of The Cybernetic Cut—the formal side of reality—to the near side—the physicodynamic (physical) side of the ravine. The scaffolding needed to build The CS Bridge consists of devices that allow instantiation of formal choices into physical recordations of those choices. This is accomplished through the construction of physical logic gates—the equivalent of Maxwell's demon's trap door. The gate can be opened or closed by agent choice at different times and in different contextual circumstances. The open or shut gate corresponds to "yes" vs. "no," "1" vs. "0." Because the gate can be opened or closed by the operator at will, we call it "configurable." It's the equivalent of an "On" or "Off" configurable switch.[42]

No physical force determines how the configurable switch is set. On a horizontal circuit board with old-fashioned binary switches, the forces of gravity and electromagnetism work equally on either possible setting of these switches. The only other forces of physics, the strong and weak nuclear forces, are also irrelevant to how configurable switches are set. Only one thing determines how they are set—choice-contingency. The deliberate, purposeful setting of a single binary configurable switch constitutes crossing The Cybernetic Cut across the CS Bridge.

Another means of crossing the CS Bridge across The Cybernetic Cut is to select physical symbol vehicles (tokens) from an alphabet of tokens available in a material symbol system. Like configurable switches, the tokens are unique physical devices. Each token is specially marked with a particular formal symbol. Scrabble tokens, for example, theoretically could be "randomly selected" (technically a self-contradictory nonsense phrase), just as configurable switches theoretically could be "randomly set." But universal empirical experience has long since taught humanity, including the scientific community, that "random selections" never produce or improve sophisticated programming function. "Garbage in, Garbage out!" Mutations cannot be distinguished from "garbage." The one and only factor that produces or improves sophisticated function is purposeful and wise choice-contingency. The specifically symbolized tokens have to be deliberately chosen from an alphabet of "physical symbol vehicles"[45,59,60] to spell a meaningful message. Similarly, configurable switches have to be deliberately set to integrate a circuit or to successfully program computational success. The essence of crossing the CS Bridge across the vast ravine of The Cybernetic Cut is *purposeful choice-contingency*.[42]

Traffic flow across The CS Bridge has thus far been observed to be one-way-only. Said Howard H. Pattee,

"The amazing property of symbols is their ability to control the lawful behavior of matter, while the laws, on the other hand, do not exert control over the symbols or their coded references."[18]

Despite many decades of life-origin science trying to bridge the gap, The Cybernetic Cut[41,42] remains untraversed except across the unidirectional CS (Configurable Switch) Bridge.[42] Formalism can be instantiated into physicality. But physicality cannot reverse the traffic flow across the CS Bridge to invade the world of formal controls. The reason is that physicality offers nothing but constraints and chance-contingency with which to attempt programming controls, computation, circuit integration, complex machine generation, algorithmic optimization, organization, and sophisticated utility of any kind. Neither chance nor necessity can steer toward "usefulness," pragmatism, or generate non-trivial formal function.

Physicality cannot choose which way to throw a horizontal binary switch knob to produce desired function. The physical environment might be able to constrain the switch knob to be thrown in a certain direction if the switch comes near a magnet, for example. But if the switch just happened to be near a magnet, no formal choice determinism would be in play that would program that switch setting for potential formal function. And if multiple switches just happened to be near the magnet, all of the metal switches would be set the same way. A program consisting of all 1's, or of all 0's, would result that could not integrate a circuit or compute anything functional. All the switches would be set to "open," or all the switches would be set to the "closed" position *by law*. Programming would be impossible. Freedom from law is necessary to program. Yet chance-contingency cannot program switch settings either. Thus physicality (chance and necessity) on the near side of The Cybernetic Cut cannot generate non-physical formal controls and regulation. The introduction of choice-contingency into physicality requires traveling across the CS Bridge. All traffic across this Configurable Switch Bridge flows in one direction only—from the non-physical formal world of abstract conceptuality, organizational specification and engineering into the physical world.

If mind were nothing more than a secretion of human physical brain, both mind and brain would have to have emanated from the far side of the Cybernetic Cut. The brain develops from a zygote containing very little protein mass and structural form. No homunculus (little phenotypic animal) exists within a zygote. Only linear digital instruction is instantiated into nucleoside selection, sequencing and epigenetic switching. Little resides in a zygote other than PI to explain the origin of brain or any other organ system. Even the non-genetic cytoplasmic PI is three-dimensionally prescriptive. Brain and mind are computed according to linear digital programming instruction. Degrees of freedom are pre-programmed to allow for end-user response to environmental challenges. But living organisms are largely controlled by prior programming. This programming includes embryological development and the most extraordinary metamorphoses. Organisms respond expediently to environmental challenges and stresses with pre-programmed end-user freedom. All known life is cybernetic. All of these controls had to be programmed at true decision nodes prior to the existence of any environmentally selectable phenotypic fitness.

"But nucleic acids and proteins are *physical* molecules," we counter argue. "Life is three-dimensional, structural and dynamically-maintained far from equilibrium in an open local environment." But the same counter argument could be said of the very physical configurable switches that exist in any computer. We do not attribute computational halting in computers to physical silicon molecules. We attribute computation to how those physical configurable switches are arbitrarily set in their dynamically-inert functionality dimension.

In addition, this Cybernetic Cut and CS Bridge existed prior to the existence of *Homo sapiens*. It is not a function, therefore, of human epistemology. It is ontological, and existed early on in life's history at the subcellular level.

2.15 The fundamental unit of Choice Determinism (CD) is the binary "decision node."

The Choice Determinism (CD) that makes prescription possible can be represented by discrete unambiguous symbols, each representing a Prescriptive Choice (PC). The most fundamental unit of prescription is a single binary purposeful choice from among two possible options. Each option must be assigned a notation, or symbol, representing *the specific choice*. Thus, paired hot and cold faucet handles present the agent with an *opportunity* to make a purposeful binary choice—a true decision node, not just a bifurcation point (fork in the road). If no choice is made, no water flows, and the term bifurcation point still applies. But if water is turned on, one of the two faucets was opened. A binary *choice* was made.

Once the choice is made, and the water comes out of the faucet, the water will be either hot or cold after 30 seconds. No uncertainty exists at all once our hand gets burned by the hot water. We can represent which of the two options were purposefully selected at this "configurable switch" (the faucet handle) with either an "H" for hot, or a "C" for cold. The recordation of an "H" represents the "certainty" of what purposeful choice was made, and the functional result. The symbol "H" is not a statistical measurement of possibilities. It is a declaration of a particular purposeful choice. Empirical experience teaches us that, under normal working conditions, the purposeful choice of "H" gives us the hot water we *want*.

The most fundamental symbols for a prescriptive choice are a "0" and a "1," usually representing the difference between choices of "on/off," "open/closed," "yes/no," "true/false," "+/-." A discrete unambiguous symbol represents a specific prescriptive choice from among multiple options (at least two, with "an excluded middle" in logic operations).

This specific choice at each bona fide binary decision node is quite different from a mere "place marker." Place markers exist to reserve a physical location at which potential choices could be held. A prescriptive choice is quite different from a "bit." A bit represents only a place-marker for a *potential* binary choice. It is not a choice! Bits can measure the number of future binary choice *opportunities*, or *the number* of binary place-markers needed or already used, but *never the specific choices themselves*.

We define "function" on the basis of what agents desire or want. Notice the word "want." This is a formal word and concept, not a physicodynamic term from physics and chemistry. The same is true for the word "work," until we bastardize the term in physics by confusing it with mere movement of mass through space. The only sensible use of the term "work" is the intuitive one. The wind blows a tumble weed along the ground. In physics we might call that "work." But that kind of "work" does not measure up to the everyday formal usage of the word and concept of useful "work." Useful work is always related to desired function. And desired function is always related to agency, never to mere inanimate physicochemical interactions.

2.16 Can programming choices be measured with fixed units of measurement?

Prescription is a product of Decision Theory, not Stochastic Theory. Deliberate decisions cannot be quantified with statistical combinatorial units of uncertainty. "Bits" are a measure of uncertainty, not prescription. Unlike additive, fixed binary units of "bits" of Shannon Uncertainty, Prescriptive Choice (PC) cannot be measured with units of fixed value. Each choice within its syntax and application is unique. Its role, within the sequence, programming and processing of choices, varies greatly. The relative worth of each individual PC lies somewhere on a variable "gray-scale" of fuzzy value relative to each computation and integrated, prescribed function. No greater example of this exists than in the sequencing of nucleotides in DNA and amino acids in proteins. Certain segments of the overall sequence are much more important than others. We cannot assign each choice of nucleotide or amino acid equal unit value.

We can summate and multiply combinatorial "bits" of Uncertainty. We can measure "fits" of Functional Sequence Complexity (FSC).[6,72-74] But we cannot measure Prescription of gray-scale meaning or function with fixed units of measure. The fact that we cannot assign the same fixed value to all units of prescriptive choice frustrates us to no end. But this curse is the very blessing of PI. It is the relative value of each prescriptive choice "unit," and of the syntax of those specific choices, that allows such relative degrees of creativity in design, architecture, engineering, art, poetry, theatre, prose, music, even altruism and ethics. We may think of these examples in terms of the humanities. But, it is also true of the fine-tuning of subcellular prescription and processing. It is incumbent upon the materialist/naturalistic scientist to explain how chance and necessity, mass and energy, could have produced such finely-tuned functional creativity.

Even "bits" of reduced uncertainty ("mutual entropy") are still a measure of uncertainty. Reduced Shannon uncertainty ($H_1 - H_2 = H_3 = R$) can never rise above a measure of negative nonspecific uncertainty (H_3). "R" may eliminate possibilities, and may even indirectly represent gained (extrinsically provided) human knowledge of that elimination. But, that gained human knowledge was not derived from the equation itself. Reduced uncertainty itself does not provide or measure positive, specific, choice-contingent Functional Information (FI). Real information provides relative certainty of exactly what *will* "work for us." Just as with controls, Functional Information (FI) is invariably the product of deliberate Choice-Contingent Causation and Control (CCCC). "Make this particular choice, not that choice."

Shannon theory has nothing to do with specific efficacious choices—the essence of Prescription and its Processing. "R" does not tell us what specific choices we should make. The facts are clear: *positive efficacious programming choices cannot be measured by combinatorial statistics!* Bits measure the number of binary choice *opportunities*, not the actual choices themselves.

The "fits" of Functional Sequence Complexity (FSC)[6,73,74] and the measurement of so-called "functional information"[4,37,75] are not measures of Prescriptive Information (PI). They are both modified measures of Shannon Uncertainty. Those measurements are educated only in the sense of what *percentage* of combinations have some degree of function. The function can be specifically named. A yes/no, empirical, after-the-fact, answer can be given as to what percentage of combinations achieve any degree of that function. But this specific information cannot be ascertained from combinatorial uncertainty measures. PI must be provided extrinsically. We must be given the information of which *particular* combinations work. The *specific* information of which particular combinations work is found nowhere within the educated bit quantifications. Neither FSC nor the Hazen, et al. calculations measure true "information." True information answers the question, "Which specific combinations are certain to work for us to achieve the function we desire?" The answer to this latter question is what constitutes Prescriptive Information (PI).[2,21,22,30,31,42]

The classic "bench-science" school of thought requires measurement before being willing to entertain that subject matter can be scientifically addressable. But this results in an extremely naïve view of science. All kinds of legitimate fields of science exist that cannot reduce their subject matter to measureable quantities. Most of the scientific method itself consists of non-quantitative axioms, ideas and generalizations. Non mathematical logic theory tends to be non-quantitative. How far would science get without logic theory? Look at astrophysics, and particularly cosmology and cosmogony. The laws of physics themselves, and measureable quantities, did not even exist until after the Big Bang. Look at sciences like archaeology and anthropology. We cannot eliminate a large number of scientific disciplines just because some aspects of their subject matter are non-quantifiable.

Specific programming choices will never be quantified by variations of measuring Shannon's combinatorial *uncertainty*. PI is a form of certainty, not uncertainty. When we buy software, we are paying for the relative certainty that the purchased software will produce the optimized computational function we paid for. If the program crashes, we want our money back. The software had "bugs" that compromised its functionality. "Bugs" are bad purposeful programming choices. We do not pay good money for probabilistic uncertainty measures, mutual entropy (reduced uncertainty) included. We want the string of symbols that represents the specific choices that have been proven to work best. That is all we care about. What we pay for is precise guaranteed Prescription. It cannot be measured with units of statistical uncertainty.

Prescriptive Information (PI) is processed through the reading, and acting upon, a string of symbols each representing a completed, specific, purposeful programming choice. Each choice is represented by a recorded symbol of instruction or pre-programmed control. The central processing unit relies upon a specific completed and recorded instruction that is known to be efficacious in realizing a functional goal. The processor is not interested in "bits" of Uncertainty or even possibility. The processor is not interested in "fits" of Functional Sequence Complexity (FSC), either. The processor is designed and engineered to perform a specific instruction which, in combination with other instructions, will produce computational success every time. The program will "halt," with the hoped-for result of computational success.

Each prescriptive choice can be symbolized, but only after a specific purposeful choice is made. The symbol of prescription must represent an actual choice, not a general alphabet or list of possible choices. Ambiguity and Uncertainty cannot specify particular choices in pursuit of utility. Options are no longer relevant to the symbolic representation of a past tense, specific, purposeful choice in Prescription theory. When programming choices fail to produce successful computations, for example, one can question whether the prescriptive choices made were the wisest.

Not all prescription is successful. But the only way successful non-trivial prescription can be produced from the world of Shannon statistical combinatorial Uncertainty is through Choice-Contingency. One cannot mix the categories of Uncertainty with Certainty. Prescription provides the certainty of which specific choices will "work for us."

Science awards Nobel Prizes for highest relative value of choices leading to the discovery and elucidation of how the objective world works. Like it or not, each of those choices do not have a fixed value. We award ingenuity and creative exploration subjectively, not mathematically. We have no choice. We cannot award the prize on the basis of fixed units of Shannon statistical combinatorial uncertainty. We award the Prize for purposeful, insightful choices at true decision nodes, some of which have much more profound value than others.

Shannon is celebrated for his prescriptive choices in transmission engineering. His achievement was to measure the statistical uncertainty of *all potential* messages with fixed units of measurement. This was exactly what was needed to deal with planning for all potential telephone calls, rather than certain ones with specific meaning. The certain phone calls with specific meaning can be produced *only* by Choice-contingent Causation and Control (CCCC), otherwise known as Choice Determinism (CD). The relative value of *certain* phone calls constitutes an almost infinite gray-scale relative worth that defies quantification.

2.17 Both PI, and the processing of PI, require Mind and Agency.

We have to stop here a minute to define both "mind" and "agency."

"Mind" is intangible thought. For thought to have meaning, that thought must be arbitrarily-chosen according to pre-defined rules. Freedom from Physicodynamic Determinism (PD) is required.

Professional philosophers, including philosophers of science, define "agents" as living beings able to "act" into and influence the cosmos. Agents can change the physical world through the unique ability to make arbitrary, yet purposeful choices.[76] Choices are distinguished from natural forces, which cause only "unthinking" or physicodynamically deterministic processes.[77] Agent choices, on the other hand, are freely made. They are not viewed as the product of physicochemical causal chains.[78] This reality generates perpetual questions relating to the mind/body problem of philosophy. Such questions will never be answered at the physical brain level. Metaphysical naturalism always reverts to attributing mind to physical brain. Mind is seen as an epiphenomenon of brain physiology: "The brain secretes thought as the liver secretes bile."

In a laughable attempt to circumvent the problem of prescription and its need for mind and purposeful choice-contingency by agents, Hazen tries to redefine an agent as a grain of sand, or a mere molecule!

"Three principles guide natural pattern formation in both biological and non-living systems:

1. patterns form from interactions of numerous individual particles, or agents, such as sand grains, molecules, cells or organisms;

2. assemblages of agents can adopt combinatorially large numbers of different configurations;

3. observed patterns emerge through the Selection *FROM AMONG* highly functional configurations. These three principles apply to numerous natural processes, including the origin of life and its subsequent evolution.[3]"

Ascribing to a molecule or grain of sand the properties of "agency" flies in the face of hundreds, if not thousands, of years of academic tradition, including not only the philosophy of science, but science itself. The scientist has never been grouped with inanimate things, let alone a solitary molecule or grain of sand. Neither a molecule nor a grain of sand desires or pursues eventual function and utility. Neither can exercise purposeful Choice-Contingent Causation and Control (CCCC: Choice Determinism [CD]). A molecule can't program. A grain of sand can't prescribe any non-trivial function. No basis for preference or Selection *FROM AMONG* isolated function over non-function exists in a prebiotic environment consisting of nothing but inanimate things. Organization doesn't emerge from physicodynamically-ordered configurations. These pronouncements, taken from a peer-reviewed science journal, aren't based on repeated observation or any prediction fulfillments. They arise from nothing but metaphysical naturalism's Freudian wish-fulfillment.

In an effort to circumvent the impossible dilemma that chance and necessity cannot organize or program, the King's English gets completely redefined in an instant to render a molecule or grain of sand a sentient, choosing, organizing "agent." That way, the absurd fanatical belief in emergence, along with the spontaneous generation of life, will sound more scientifically plausible.

Combinatorial complexity has absolutely nothing to do with function.[6,29,30,61,62] The greatest degree of complexity is randomness. Since when has randomness ever generated computational success?[29,62,79-81] No basis for selection exists in a prebiotic environment.[21,22,31,32,42,71,82-86] Molecular evolution will be examined in great detail throughout Section 3 of this book. It is nothing less than a pipe dream, with zero rational or empirical support. It is not even falsifiable. "Emergence" is found under proper scientific skepticism to manifest far less plausibility than what naturalistic science typically calls "pure religion." Whatever we call it, the notion of spontaneous "emergence" of such sophisticated organization and integrative function remains shameful "blind belief." At best, it is science fiction rather than science.

The Gaia philosophic movement contributed greatly to the view that inanimate earth was "alive," and "a living organism." The earth is no more alive than any other planet or moon.

Why is the need for animate agency so important for science to acknowledge? First, many scientific disciplines such as archaeology and anthropology have to deal directly with the fact of agency as the very subject of their scientific investigation. Second, all scientists themselves are agents. Science is an epistemological system of human minds. The scientific method is a composite of innumerable nonphysical formalisms (e.g., mathematics, language, conceptual categorization, logic theory, representational symbol systems, and computation) that can only be practiced by agents. The philosophic naturalist simply cannot avoid or deny the fact of agency as a part of "the real world." Agency itself is a legitimate subject of scientific investigation (e.g., anthropology; psychology; archeology).[87] But, most important of all, biology has no choice but to acknowledge that life is cybernetic, and that cybernetics (formal control) requires Choice Determinism (CD) rather than Physicodynamic Determinism (PD).

Mind and agency involve some major unique attributes that must be considered:

2.17.1 Self-awareness

Self-awareness is the conscious awareness that one exists as an individual person or organism. Self is perceived differently from "other." "Other" includes the inanimate environment and other living individuals within the environment.[88] The capacity for introspection also seems to be a major component of self-awareness.[89]

Sub-human animals' self-awareness is more limited than that of humans. Introspection, especially, seems minimal apart from *Homo sapiens*.

Agents are typically self-aware of their goals and activity in pursuit of their utilitarian desires.[77] Agency can sometimes operate subconsciously, without apparent awareness of voluntary choice behavior. But even subconscious preferences involve motivation, desires and beliefs. Agents intentionally implement direct controls, steering, and pursuits of fulfillment of their desires and perceived needs.[90]

The naiveté of materialism/physicalism becomes much more apparent when the mind/body problem is reduced to simpler models. The simplest model is found in the field of ProtoBioCybernetics. This specialty focuses on how *control and regulation* first arose in an inanimate environment ruled by nothing but chance and necessity, mass and energy. How did the very first gene arise? How was the first gene processed? How were the very first epigenetic switches set that controlled that gene's expression?[91]

No brain exists yet in a prebiotic environment to secrete thought, let alone purposeful choices. Yet, programming choices had to have been made in order to generate the first subcellular controls and programs. Even a primordial cell depended upon programmed controls that could be passed on to offspring through cell division. Otherwise, the wheel would have to be re-invented with every new protocell that happenstantially formed. Not even evolution could have occurred without the heritability of programmed instructions that could mutate at the genomic level.

2.17.2 Valuation, Desire and Motivation

Inanimate nature has no perception of "usefulness." No preference for utility exists. Non-function is just as good as function to an inanimate environment. The environment does not pursue potential function, and does not value it. Prebiotic nature had no goals or schemes to achieve pragmatic aims.

Only minds and agency place relative value on entities. Motivation follows.

2.17.3 Intention and pursuit of functionality

Intention is an aim, plan, target, aspiration, ambition, scheme, objective or determination to act with purpose to achieve some goal.[92,93] Intention is unique to "agency." Agents are deliberate choosers and pursuers of utility and self-interest.[94]

Sub-human animals can also manifest intention. One of the most stunning examples is an ant eater using a stalk to insert into an ant hole in order to ingest ants that walk out of the ant hole onto the stalk. Prokaryotes can "choose" to approach nutrients and avoid noxious stimuli. In a very broad sense, even plants can intend to pursue sunlight (phototropism) or water (hydrotropism). Such "intention," however, is often seen as being preprogrammed rather than free voluntary agency. Some try to extend this perspective to embody all of human behavior, at least until they wish to take personal credit for their superior academic research. Then, suddenly, their agency is celebrated over their dogmatic Skinnerian behaviorism.

No inanimate object has ever been observed to manifest intention or purpose. Intention is unique to already programmed, already-living organisms, usually those with some semblance of a central nervous system.

2.17.4 The choice-contingency needed to achieve formal utility.

Empirically, choice-contingency has invariably been found to be a function of agency and mind. Weber attempts to summarize the contrast between mental causation and indeterminism while dealing with the mind-body problem.[95] Thus far, very little progress has been made in trying to reduce mind to physical brain. The primary reason is the inability of chance and necessity models to generate and explain the phenomenon of steering events toward non-trivial utility.

Non-trivial function cannot be programmed, prescribed or achieved without choice-contingency. Chance (e.g., coin flips, whether fair or unfair) cannot program sophisticated function. Law and initial condition constraints cannot program either. Wise programing choices must be made to achieve non-trivial computational success. Only mind and agency can purposefully choose from among real options.

2.17.5 Knowledge of/about ontological being (objective reality)

Another characteristic of agency is knowledge of or about a presumed external objective world. Knowledge is not an end in itself. One of science's most fundamental beliefs is that scientists, as agents, can gain a progressively more accurate understanding of ourselves and our objectively existent environment. Our knowledge of the world can grow in correspondence to the actual real world.

Knowledge can be mistaken. But there is an objective standard with which our knowledge can be compared through double-blind studies, independent teams of investigators all doing the same basic experiments, scientific debates at conferences, letters to the editor, and comparative papers in the literature. Objective, ontological Truth is presupposed in science despite a healthy acknowledgement of humans "epistemological problem."

3. How did primordial prescription, and its processing, arise in a prebiotic environment?

Science is not just about facts. Science is about "How?" How do things come into existence in a cause-and-effect world? What is the mechanism of causation for the effects of genetics, genomics, and epigenomics?[96] It is absolutely mind-boggling how so many origin-of-life books and papers could have been published over the last five decades that completely ignore, or side step, by far the most important questions of life origin. We could list at least a thousand references here, but will cite just a few representative examples of what can only be considered obfuscation (almost deliberate sweeping under the rug) of the most important questions of life origin.[3,4,77,97-222]

Clive Trotman should be given credit for at least some healthy degree of naturalistic honesty in his book.[223] Pier Luigi Luisi in recent years has also been more open and honest about difficult questions.[224-226] Before any prescriptive genetic information can be duplicated and "vary," those instructions must first exist.[46-48] How did the very first Prescriptive Information (PI) get programmed and recorded into a representational material symbol system in a prebiotic environment? How did a processing mechanism simultaneously arise, and the machinery needed to interpret and put into effect those instructions?

PI instructs or provides previously-made purposeful choices that program non-trivial function. Does this hold true for life? All known life is cybernetic (highly controlled and regulated). No question is more central to life-origin research than the origin of prescription and the processing of that sophisticated utility. Is it our serious contention that life doesn't need to follow all of the same rules that any other form of Prescription and Processing (P & P) has to follow? On what basis could we defend that contention?

3.1 The chance and necessity of nature are blind to function and "usefulness."

Chance has no perception of functionality, or anything else. Chance has no interest in utility. Chance represents whatever probabilistic combination happens to unfold.

Random molecular motion and heat agitation possess no capabilities to organize, engineer, build sophisticated structures, or program computational programs. In fact, chance is not a cause of any effects. Chance is merely a mathematical, descriptive (not prescriptive), formal mental construct. Chance has no goal of achieving usefulness. Chance manifests no pursuit of *useful* work. Chance cannot design or engineer machines or systems. It can't program anything. Chance can't bring a non-trivial computation to successful halting.

But, "necessity" (law-like cause-and-effect determinism) has no more interest in function than "chance." Fixed, unimaginative, redundant, high-probability natural law operates without reference to whether any physical interaction might be "useful." Physical interactions proceed as a blind chain of cause and effect. Initial conditions are acted upon by non-sentient laws of physics and chemistry.

Force fields could care less whether anything "works." Law generates no freedom from physicodynamic cause-and-effect determinism. It is therefore impossible to program Decision Nodes under circumstances of physical determinism. Prescriptive Information (PI) cannot possibly be generated by fixed law. The first requirement, even of Shannon "information" (which is not really information), is combinatorial uncertainty. A physical matrix or medium must have combinatorial uncertainty (chance-contingency) into which purposeful choices could be instantiated *if* any bona fide PI were to be stored.

Law has no goal of achieving usefulness. Law manifests no pursuit of *useful* work. Law cannot design or engineer machines or systems. Law can't program. Law can't drive a non-trivial computation to successful halting. Zero "pressure" exists from differential survival to prescribe or process creative new utility. Shannon bits measure only combinatorial Uncertainty. There is virtually no uncertainty (theoretically, at least) in any law-like behavior. The probability of law-like behavior approximates 1.0. That calculates to approximately 0 bits of Uncertainty (no bona fide information *potential*). A few bits of Uncertainty are found in standard deviation curves surrounding law-like determinism. But such noise does not program computation.

You've heard the famous Tom Hanks movie line, "There's no crying in baseball!" Well, "There's no 'usefulness' in physics!" Physics and chemistry do not value or pursue pragmatic considerations. The use of the word "work" in physics was bastardized from the formal concept of work, the same way the word "information" was bastardized from intuitive, semantic information into Shannon's "Uncertainty" theory. Shannon himself objected to referring to his transmission engineering work as "information theory."[67] He pointed out early in his first paper that it had nothing to do with intuitive, semantic information.[67]

When one chooses to metaphysically believe that "chance and necessity (law)" are all there is, nothing remains as a possible explanatory mechanism for P & P. If inanimate nature is utterly blind to utility, Chance and Necessity have no programming or processing capabilities. This is also true of any *combination* of chance and necessity. No synergism exists between the two. Neither mass nor energy can explain sophisticated function. Mass/energy inter-conversions and phase changes cannot explain sophisticated function either. Physicodynamics and physicochemical interactions/reactions are oblivious to formal, abstract concepts such as "Usefulness." They are also utterly blind and indifferent to notions of "value." A lifeless environment cannot exercise desire for utility. Physics and chemistry play no role in *generating* "functionality." The physics definition of work is completely unrelated to the intuitive, semantic, everyday connotation of "useful work." (See Section 3.18)

The chance and necessity of nature do not, and cannot, program computational success. Mass and energy cannot pursue utility.

Any pragmatist should strongly object to naturalists trying to use the physics definition of work to refer to intuitive "useful work." The physics definition of "work" is utterly blind to utility, and cannot possibly address or explain "useful work."

Freedom from Necessity is required to choose at bona fide decision nodes. If choices were forced by law, they would not be true choices. They would be mere constraints. Any program written by constraints or law would consist either of all 0's or of all 1's. No computational program has ever been written "by law." It could contain no prescriptive information because it would not arise out of contingency. Information instantiation into any physical medium is impossible without freedom from law. Decision nodes require contingency and uncertainty. A decision node provides an opportunity to convert Chance-Contingency into Choice-Contingency.

It is impossible to program Decision Nodes under circumstances of physical determinism. Prescriptive Information (PI) cannot possibly be generated by fixed law. The first requirement of "information" is combinatorial uncertainty in the matrix or physical medium into which purposeful choices have to be instantiated. This is what Shannon bits measure: Uncertainty. There is virtually no uncertainty (theoretically, at least) in any law-like behavior. Of course, as mentioned above, there are standard deviation kinds of statistical indefiniteness with any law-like behavior. But, that statistical variation has no cybernetic benefit. It is more like "noise pollution" of fixed law determinism, rather than noise pollution of any meaningful message or instruction in the Shannon channel.

To achieve freedom from law in physical media requires a specially designed configurable switch that can only be set by formal purposeful choice. No physical force can set the configurable switch. An agent must push or turn the knob in one direction or another. The same is true with Material Symbol Systems (MSS).[44,45] Freely resortable tokens ("physical symbol vehicles") and their sequences must be arbitrarily chosen. No force of nature can do this. Examples of physical tokens being used in a Material Symbol System (MSS) are the lettered blocks of wood used in the game called Scrabble. Another example of tokens used in an MSS includes nucleotides in DNA.

We know from biology that intracellular and intercellular programming are quite real. We know that this programming not only predates *Homo sapiens* and their consciousness, but that it predates metazoans (multi-cellular organisms). It predates eukaryotes (nucleated cells with more sophisticated organelles). Even prokaryotes, whether bacteria or archaea, are more ingeniously organized and programmed than the world's finest main-frame computers.

There can be no denial that life is genetically, genomically and epigenomically programmed, controlled and tightly regulated at every turn. All known life is cybernetic. If we cannot deny that life is cybernetic, we cannot deny that life is governed by formalisms. Formalisms involve purposeful choices and formal rules rather than forced physical laws.

Two things are certain:

1) Programming choices cannot be optimized for sophisticated functionality if they are made by coin flips!

2) The programming choices will not be optimized for sophisticated functionality if they are made by physical law, which precludes functional programming altogether!

Before life is ever mentioned, we need to take to task any physicalist who is trying to extract PI, Instructions, Functional steering, or programming of any computational function out of fixed, invariant Physical Law (or the laws of motion). Mass/energy interactions, with their cause and effect determinism and slight statistical variance, cannot program anything. It is a *logical* impossibility for chance and/or necessity to program any cybernetic function.

Naturalists have no choice but to turn to quantum quackery to try to explain the undeniable, repeatedly observable, FACT of cybernetics in reality. But, long before muddying the water with any biological questions, we need to expose the fallacy of attributing *any* form of cybernetics to stochastic events. Quantum events are probabilistic. They cannot explain Decision Theory programming of sophisticated functionality.

We also need to take to task any physicalist who tries to program formal functionality using nothing but "possibility." Why do we laugh at the Yogi Berra-ism, "When you come to a fork in the road, take it!" Why is that funny? We laugh because we know that merely acknowledging a "bifurcation point" tells us nothing about *which* fork we should take. All acknowledging "possibility" does is to acknowledge contingency and uncertainty. "Possibility" doesn't program a single computational function. "Possibility" provides no Prescriptive Information (instruction for how to achieve any pragmatic benefit).

Yes, it's possible that taking one of the forks in the road will get us to our destination more efficiently. But, has that fact ever been at issue? Does anyone doubt that different courses of action are possible? Or that one choice might be better than another? But what does that information tell us about *which* choice to make?

When we buy new software, we don't spend our money on "possibility." We spend our money on already-prescribed choices (from among many possible options). We pay for *specific* choices that are already proven to optimize the computational function we want.

So, why are we SO in love with "possibility?" "Possibility" is all we have with which to work and appeal to when trying to explain cybernetics in a purely physicalist worldview. All that really matters is the answer to the question, "*Which* possibility works best?" When the specific question of "which?" is ignored, or swept under the rug, the "possibility" we worship fails to provide a single efficacious choice. "Possibility" alone boils down to nothing but noise and combinatorial uncertainty.

Everything really interesting in life can be traced back not just to combinatorial possibility, but to *which specific wise choices* at bona fide decision nodes accomplish sophisticated utility. Physicodynamics makes no purposeful decision-node choices. Materialism/Physicalism/Naturalism is impotent—utterly devoid of any hint of explanation or plausible model for how cybernetic usefulness was achieved by nature.

Life in the final analysis is not about quantum stochastic possibility. It is about the unique Decision Theory choices that alone generate integrated circuits and make organized cooperative metabolic schemes possible. Neither chance nor necessity can explain the most impressive empirical data known to science: the cybernetic nature of all known life. This is true at life's subcellular level, and at its intercellular organ, organ system, and organismal levels.

3.2 No yet-to-be discovered law could possibly prescribe sophisticated function.

Above we saw that not only can chance not prescribe sophisticated function, Law cannot prescribe it either.

Prescription requires wise choice from among real options (choice-contingency). Law-like constraints eliminate options. Law favors only fixed, redundant, unimaginative constancy. The very essence of law is extremely high-probability redundancy. This is why parsimony is possible in laws. High probability makes programming freedom impossible.

If law cannot possibly program, that fact would also apply to all yet-to-be discovered laws. The *last* thing that could explain programming of initial bio-organization and genetics in a prebiotic environment would be some new yet-to-be-discovered law of physics or chemistry. Law is poisonous to prescription because it precludes programming choices. Programming requires freedom from law, whether that law is already discovered, or yet-to-be discovered. Law is law. The only reason any law could be formulated is that the probability of what that law describes approaches 1.0. The principle of Ockham's razor arises out of the simplicity provided by such high probability of law-like interactions.

Cybernetic process can only be prescribed by "choice with intent" at bona fide decision nodes. Chance and necessity cannot program configurable switch-settings into highly integrated circuits. Computational success is a formal phenomenon, not a physicodynamic effect. Mass and energy cannot select tokens in a material symbol system to "spell" meaningful, coded instructions. Inanimate molecules cannot organize themselves into long-term sophisticated architectural structures and machines. The Second Law tendency toward disorganization would have precluded sustained primordial spontaneous "self-organization." Chemical interactions cannot generate processing schemes and equipment.

3.3 The Universal Selection Dichotomy (USD)

Three fundamental categories of reality exist: Chance, Necessity and Selection.[22] Two fundamentally different categories of Selection also exist:

1) Selection *FROM AMONG*

2) Selection *FOR (in pursuit of)*

The distinction between these two classifications of Selection constitutes *The Universal Selection Dichotomy (USD)*.

3.3.1 Selection *FROM AMONG*

Selection *FROM AMONG* requires existing options or alternatives from which to choose. In the case of evolution, the environment "Selects" the fittest living organisms *FROM AMONG* already-existing phenotypes. These organisms are already programmed genomically and epigenomically when neoDarwinian natural selection takes place. Their superior programming is ultimately the reason for their superior phenotypic fitness and eventual "selection."[23,24,26-28] The selection does not take place at the genetic or epigenetic molecular level (The GS Principle.[71]). Selection takes place only secondarily, at the phenotypic level.[227] Natural selection, therefore, is an example of Selection *FROM AMONG*, never Selection *FOR (in pursuit of)* at the programming level.

When we purchase artificial computer devices, we are choosing from among already existing hardwares and firmwares. When we purchase applications, we are choosing from among already existing softwares. Consumers usually know little to nothing about computer-science cybernetics. They simply select from among whatever hardware, firmware and software options exist on the store shelf. Like the environment, they are ignorant of programming concept, architecture, processing and execution. Consumers play no role in programming computational success. Their selection is based solely on cost and reputation of the efficacy and efficiency of existing hardware and/or applications. The consumer prefers and Selects *FROM AMONG already existing "fitness."* The same is true of environmental selection.

3.3.2 Selection *FOR* (in pursuit of)

Selection *FOR (in pursuit of)* represents choices made at bona fide decision nodes stemming from the desire for, and the goal of obtaining, pragmatic utility.[41,42] What makes Selection *FOR (in pursuit of)* unique is that the choices must be made prior to the existence of any relative function and value. No utilitarian results exist at the time of selection to compare, prefer, or choose from among. Evolution plays no role whatever in programming choices.

Selection *FOR (in pursuit of)* generates novelty. It is creative. Selection *FOR (in pursuit of)* programs computational halting and pragmatic success to produce a first-time, formally useful product.

Non-trivial "usefulness" is an abstract, conceptual phenomenon and goal. While the cybernetic product may be physical, its origin depended entirely upon formal intent, programming choices, integrated logic gate assignments, and astute, cooperative, configurable switch-settings. Utility normally has to be "envisioned" in its developmental stage, prior to the realization of any goal.

Computer programming is an example of Selection *FOR (in pursuit of)*. The programmer must make purposeful choices not only in anticipation of, but to the end of, actually bringing into existence a functional computational program. The choices that generate computational success must be made in advance, individually, to generate not-yet-existent computational "halting." Successful computational function is desired, sought after, schemed for, organized, and actually produced by purposeful choices at each decision node. Halting can only be verified empirically, after the fact of its programming choices have been made (the famous halting problem[1]).

When Maxwell's demon[30,228-241] selects a trap door setting "in order to concentrate the faster moving, hotter atoms on one side of the partition," he is Selecting "*FOR*, or in pursuit of, potential function (creating a usable energy potential)." **See Figure 2 this Section, 3.3.2**. The demon's operation of the trap door represents bona fide choice-contingency at the programming level, prior to the realization of any utility. No heat engine exists yet to make use of the energy potential.

Figure 2. Maxwell's Demon has to purposefully choose when to open and close the trap door to accomplish his goal. Because the gas molecules are inert and ideal, no physicodynamic mechanism seems apparent to explain spontaneous self-ordering, let alone formal self-organization, to the utilitarian end of creating a non-trivial heat engine. (Used with permission from: "Moving 'far from equilibrium' in a prebiotic environment: The role of Maxwell's Demon in life origin." David L. Abel, In *Genesis - In the Beginning: Precursors of Life, Chemical Models and Early Biological Evolution*, Seckbach, J.Gordon, R., Eds. Springer: Dordrecht.)

The trap door is a binary decision node, a binary configurable switch, which becomes a true "logic gate." In Maxwell's thought experiment, the trap door is either open or shut, with a theoretically excluded middle. The demon's purposeful choice and intent is to set the logic gate to either "open" or "closed." This alone promotes a local and sustained anti-equilibrium state. The only way such a sustained anti-equilibrium state can be achieved is through Selection *FOR (in pursuit of)* the goal of opposing the Second Law.

Formal organization and purposeful Choice-Contingent Causation and Control (CCCC)[83] alone outsmart the Second Law tendency, temporarily and locally. In the absence of continuing wise choices, the Second Law will invariably win out, with progression back toward disorganization. Only purposeful choice-contingency organizes the energy potential into a *useful work* potential.

The demon's trap door itself is physical. But, the choice of when to open and close the trap door is *not* physical. It is formal, as formal as logic theory, language, and mathematics. These two criteria constitute the unique nature of any "configurable switch." The switch is physical. But the choice of how to set that switch is always abstract, conceptual, and non-physical. **See Figure 3 this Section, 3.3.2**, which depicts the necessity of the demon's purposeful trap door choices to achieve formal organization.

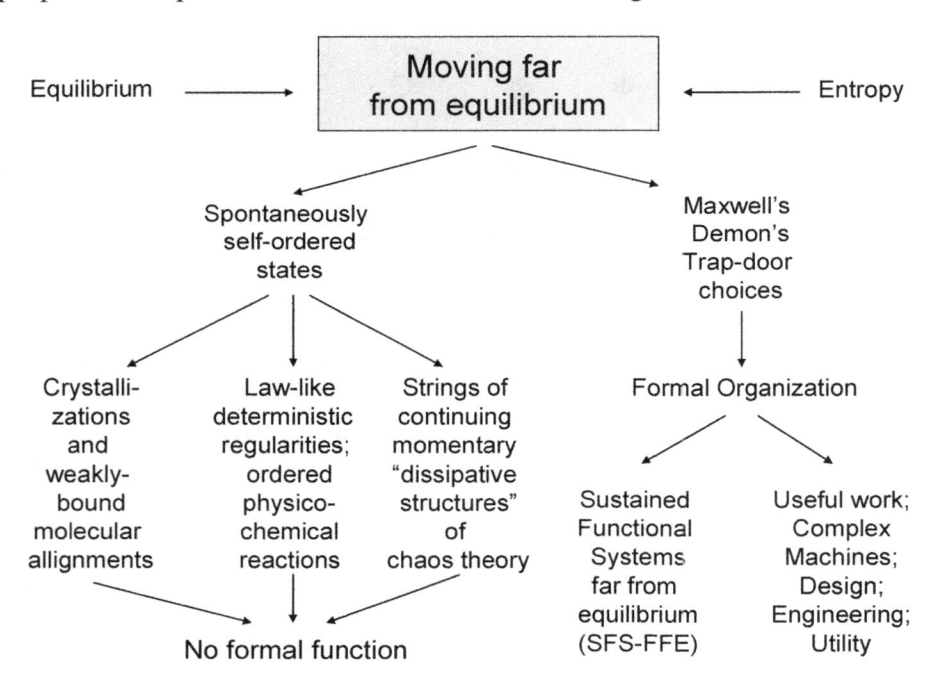

```
                          ┌─────────────────┐
        Equilibrium ─────▶│  Moving far     │◀───── Entropy
                          │ from equilibrium│
                          └─────────────────┘
                           ↙               ↘
                Spontaneously            Maxwell's
                 self-ordered             Demon's
                    states               Trap-door
                                          choices
               ↙      ↓      ↘               ↓
         Crystalli- Law-like  Strings of   Formal Organization
          zations  determin-  continuing     ↙         ↘
            and    istic      momentary
           weakly- regular-   "dissipative Sustained    Useful work;
           bound   ities;     structures"  Functional   Complex
          molecular ordered   of           Systems      Machines;
          allign-  physico-   chaos theory far from     Design;
           ments   chemical                equilibrium  Engineering;
                   reactions               (SFS-FFE)    Utility

              ↘       ↓       ↙
              No formal function
```

Figure 3. Diagram showing why "order" and merely moving "far from equilibrium" are not synonymous with organization, function or life. (Used with permission from: "Moving 'far from equilibrium' in a prebiotic environment: The role of Maxwell's Demon in life origin." David L. Abel, In *Genesis - In the Beginning: Precursors of Life, Chemical Models and Early Biological Evolution,* Seckbach, J. Gordon, R., Eds. Springer: Dordrecht.)

Formal means choice-contingent, or purposefully Choice-Determined (CD) (See Section 2.12). Formalisms are not dependent upon the laws of physics and chemistry.[19] But, they do not violate those laws, either. The configurable switch first provides "contingency" because it can be set to either binary position, open or closed, despite cause-and-effect physical-law determinism.[22,70,79] But upon the setting of that configurable switch, it acquires the additional characteristic of having been Choice-Determined (CD) by formal Choice-Contingent Causation and Control (CCCC).[41,42] The choice is both purposeful and free from physicodynamic determinism. Energy may be required to actually set the switch, but the choice of which direction to push the knob, or slide the physical trap door (open or closed), is purely formal.[19]

Configurable switches *can* be set randomly. But zero empirical evidence exists in the history of science of randomness ever having produced a non-trivial integrated circuit. No sophisticated program was ever written by coin flips, whether the coin was "fair" or "weighted".[29-31,62] Weighted coin flips only reduce Shannon Uncertainty (possibilities; contingency) and therefore reduce information *potential*.[6] Truly random (fair) coin flips offer more contingency. But chance-contingency can't program anything, let alone the ingenious computational successes we observe in the simplest living organisms.[85,86]

Rats improve their exit time from a maze only by memorizing individual efficacious choices from among real options at successive true decision nodes. These choices must be made in pursuit of the *future* function of eventually escaping the maze. No existing function exists at the time that each individual decision-node choice must be made. Selection is *FOR (in pursuit of)* potential function, not *FROM AMONG* existing function.

All known life is cybernetic. Life exists only though the prescription of future (not yet existent) integrative bio-function.

3.4 Natural Selection

The very loose extension of efficacious "choice/decision/selection" to "natural selection" provides the very foundation of evolutionary theory. "Natural selection" cannot possibly survive as a model or theory apart from "selection." But a serious problem exists in the theoretical application of "selection" to "natural selection." No existing function exists for the environment to prefer at the time that each required genetic decision-node selection is made. Nucleotides must be polymerized with rigid bonds in a certain functional sequence prior to the realization of secondary structure and three-dimensional folding into ribozymes. Transcriptive editing and translative utility of mRNAs must precede the building of any phenotypic living organism. Any function of nucleotide sequencing is eventual, not immediate. Yet the prescriptive sequencing must be consummated with rigid 3'5' phospho-diester bonds at the time of polymerization (at the time each nucleoside token is chosen and sequenced as coded instructions).

The same is true of proteins. Polyamino acids polymerize with rigid bonds prior to their folding into functional shapes and machines. The minimum Gibbs free energy that determines folding and function of proteins is prescribed by primary structure (sequencing) of amino acids prior to any folding. It is the primary structure that ultimately determines what the minimum Gibbs free energy will be. But, even before that, supposedly redundant triplet-nucleotide coding, using a superimposed hexamer coding language, prescribes specific protein-folding at the back end of the ribosome. This is accomplished through Translational Pausing (TP). TP determines folding before any three-dimensional function is ever realized or becomes selectable.[254] Other unrelated protein chaperones also help prescribe protein folding into the correct, needed functional molecular machine. Natural selection is irrelevant at this point. Such programming is Selection *FOR* (in pursuit of), not Selection *FROM AMONG* (from among already existing molecular machines).

Natural selection (NS) cannot select *for potential function*. This is doubly true at the genetic prescriptive level of symbol system and Hamming block code[242] (triplet codon) use (The GS Principle).[68,71,83] Yet, Selection *FOR (in pursuit of)* is exactly what is required for any type of programming, and for the processing of that programming.

NS can only select *from among existing functions*, and it can only do that in the form of favoring the best already-programmed, already-living phenotypic organisms. NS never selects for isolated functions. The inanimate environment could care less whether anything functions. It has no pragmatic preferences. And the environment cannot Select *FOR not-yet-existent function,* especially, such as we observe with any programming or engineering type of enterprise.

Physicalists need to explain how formal organization could have arisen out of chance and necessity. How did functional steering arise? How did mass and energy, chance and necessity first prescribe instructions, and process them?[2,5,29] Programming doesn't just spontaneously arise out of randomness. Cybernetics cannot arise from fixed, invariant, physical law either.[243] Non-trivial formal function of any kind is impossible without selection, specifically Selection *FOR (in pursuit of)* non-trivial function. The origin of neither genomics nor epigenomics can be explained by "selection pressure's" Selection *FROM AMONG* already existing phenotypic organisms. In section 3.6 we will show that naturalistic selection "pressure" simply does not exist in the real world. Selection "pressure" is a pipe dream.

Life is controlled and highly regulated. Life is fully dependent upon genomic and epigenomic programming. No Choice-Contingent Causation and Control (CCCC) or Choice Determinism (CD) at bona fide decision nodes has ever been observed to emanate from micro- or macro-evolution.[41,61] For sophisticated Prescriptive Information (PI) and its processing to increase requires CCCC (CD) at bona fide decision nodes.

Section 3: How did primordial prescription, and its processing, arise in a prebiotic environment?

To move from one level of programming to a more sophisticated level of programming in macroevolution requires more ingenious prescription. An increase in complex conceptual organization, integration of component parts, and cooperative function must be programmed. Prescription of translation from a polycodon language to a polyamino acid language is abstract. This Prescriptive Information (PI) must also be accompanied with functional processing equipment. Formal rules, not laws, must be voluntarily followed. All of these formal functions contribute to the larger programming goal resulting in "metabolism." The only problem is, evolution has no such goal.

The environment, of course, affects the genome, but almost always adversely. Any hint of benefit is trivial. The well-worn malaria-resistance example of sickle cell anemia doesn't even measure up to a trivial improvement in genomic or epigenomic *programming*. It's just a unique phenotypic reality that crenated red blood cells happen not to be preferred by the host *or* parasite. We cite this tired example out of desperation for lack of any real evidence of improved programming from mutations. At best, most mutations are neutral, and therefore unselectable, by natural selection.[244] Any sickle cell anemia sufferer will attest that they would gladly cash in their sickle cell anemia curse for increased malaria susceptibility. The mutation causes them to live in lifelong pain and exercise intolerance. They die young.

Mutations produce noise pollution in the Shannon channel, not superior new programming. Mutations, whether random or nonrandom, have never been observed to generate the kind of sophisticated new programming required for a macro evolutionary step-up to a single higher level family.

In current life, polyamino acid sequencing is prescribed by codon sequence, not by amino acid free-energy bonding preferences. If, in a Peptide World, bonding forces determined sequencing, the high degree of self-ordering would have produced low-informational, monotonous, redundant sequences of the same few amino acids. The peptides would have lacked the many specific functionalities needed even for the simplest of protometabolism. All spontaneously formed peptides would have tended to be the same. This would have restricted the large sample space of three-dimensional shapes that is invariably cited as being needed for molecular evolution.[245,246] A short period of time existed between earth's cooling (4.0 to 3.9 Ga) and the proposed 3.7 billion-year-age of life.[247,248] Life could not have evolved from a peptide world out of such a highly self-ordered minimal sample space. The nearly ideal nature of genetic coding could only have arisen out of freedom from law via choice-contingency. Programming selections for function must be free from high constraint coercion.[249] This programming of informational biopolymers requires not only Selection *FROM AMONG* each nucleoside, but, like language, Selection *FROM AMONG* unique *sequencing* of those nucleoside choices prior to any function. Programs must be written prior to computational success.

The *syntax* of nucleoside choices can also be edited (e.g., with alternative splicing). Segments can also be switched on and off, a different kind of configurable switch (epigenomic decision nodes/logic gates such as cytosine methylations). But the fundamental linear digital programming principle, using a system employing sequences of representational symbols, remains determinative of ultimate function even in the post ENCODE[250-252] molecular biological paradigm.

Any differential availability of nucleotides in a primordial soup would have produced weighted means in the formation of stochastic ensembles. This would only have reduced the number of bits of Shannon uncertainty and complexity. Information retention in the RNA matrix of a forming RNA World would have been severely restricted by both the limited availability of certain nucleosides, and/or by physicochemically preferred nucleoside-binding energies. The key to instantiation of prescriptive information into nucleic acid sequences is the fact that polymerization of nucleosides corresponds to freely configurable switch settings. Physical laws do not determine sequencing.

Base-pairing has nothing to do with sequencing of the template. Templating is usually highly ordered rather than contingent. Adsorption of nucleosides onto montmorillonite clay tends to produce homopolymers of all the same nucleoside (e.g., adenosine).[180] These contain no Shannon uncertainty. They are almost completely devoid of information potential. Homopolymers have nothing to do with prescription of function. Only the replication of existing informational sequences results in functional prescription. Source code is truly *controlled* by symbol selection and by encryption/decryption *rules, not laws and constraints*. The fixed laws of nature cannot explain formal instructions or coding function.

Genetics is a representational Material Symbol System (MSS).[43] This MSS could only have been generated by the purposeful choice of tokens (physical symbol vehicles) from among real options (four possible nucleosides at each new decision-node locus in the forming biopolymer). Life's major decision nodes are organized into syntactical polymerization options. The Selection *FROM AMONG* each nucleoside to be polymerized next corresponds to a quaternary (four-way) configurable switch-setting.

Methylation of cytosine is another example of a configurable switch-setting, or logic gate programming decision. Epigenetics could only have been generated from wise conceptual choices of configurable switch-settings. Chance (e.g., heat agitation) and necessity (laws and local constraints) could not possibly have generated such sophisticated genomic or epigenomic controls. Controls are very different from mere constraints.[61] Rules and controls steer physical interactions towards formal utility. Laws and constraints do not.[22,29,31,41,42,79]

Two categories of Natural Selection (NS) must be examined with reference to the Universal Selection Dichotomy [USD: Selection *FROM AMONG* vs. Selection *FOR* (in pursuit of)]:

1) Molecular evolution

2) Post-biotic evolution, both micro- and macro- evolution

3.4.1 Which of the two categories of the USD might have been exercised by molecular evolution?

Here we are most interested in the role of selection in abiogenesis (life origin).

How was each nucleotide selected for *potential* bio-function in a prebiotic environment? The answer of "natural selection" is totally unacceptable. Natural selection cannot operate in a prebiotic environment. Natural selection is nothing more than the differential survival and reproduction of *already-programmed, already-living* phenotypic organisms. Molecular evolution provides no basis for selection other than differential molecular stability and mutual- or auto-catalysis. In Section 3.5.1 and 3.5.2 we shall examine each: 1) molecular stability and 2) mutual- or auto-catalysis.

Three conditions would have obtained in a prebiotic environment:

1) No prior phenotypic living organisms existed from which to "choose" (select on the basis of relative fitness in a given environment)

2) No sensitivity to, or even ability to detect, stand-alone utilitarian worth existed in a prebiotic environment.

3) No desire for or preference of function (fitness) over non-function existed. Any physicochemical reaction was just as "good" as any other.

Given consistently-held naturalistic metaphysical presuppositions, these were the conditions under which abiogenesis had to have occurred.

Obviously, we cannot appeal to Selection *FROM AMONG* [Natural Selection; "Selection pressure"] as an abiogenic mechanism. An array of primitive, already-living organisms would have been needed for the fittest to differentially survive and reproduce. There were no organisms in a prebiotic environment.

3.4.2 Which of the two categories of the USD would have been exercised by post-biotic evolution?

The environment often indirectly "prefers" one small group of organisms over others in the same niche. Post-biotic evolution models all begin with a tremendous amount of already existent genetic, genomic and epigenomic programming. No explanation is ever provided for how any initial instructions arose. Those initial instructions, then, supposedly are to be spontaneously improved upon by the same chance and necessity alone that allegedly programmed them.

Duplication plus variation models should first address the question, "Duplication of what?" Models that seek to answer that question using nothing but statistical Shannon models of combinatorial Uncertainty fail miserably. Prescriptive Information (PI)[2,5] (instructions and programming), and the processing of that PI, originate only from Decision Theory, not Stochastic Theory.[6,7,29,39,62]

Selection *FROM AMONG* has nothing whatever to do with the original programming of each new genotype. The vast majority of an organism's configurable switches have already been set prior to natural selection taking place. The very reason one living organism is more fit than another is its superior genetic programing and its epigenomic control of genetic processing. Their circuits are already integrated; their computations are already halting; their structure is already organized; and they are already living before environmental selection can take place.

The phenotype is an effect, not a cause. Each new more sophisticated genotype prescribes and processes, with the help of additional subcellular machinery, the effect of a higher-level phenotypic organism.

Natural selection is nothing more than the differential survival and reproduction of the best already-programmed, already-living phenotypic organisms. Evolution is Selection *FROM AMONG* populations of living organisms. It is not Selection *FOR (in pursuit of)*. The environment and evolution have no aims or goals. Evolution cannot program choices of nucleosides, their sequencing, or their methylations switches.

When an epigenetic switch is flipped (e.g. methylation of a certain cytosine nucleotide), functionality can immediately change. Thus, we might think that we can equate the selectable phenotypic change (e.g. superior fitness) with Selection *FROM AMONG* methylation vs. no methylation. How easily we forget how many different component parts, biochemical pathways, integrated circuits, messaging and transport systems, and structural organizations have to be in place, fully organized and operational for a simple methylation to have any effect, let alone selective benefit. The effect of cytosine methylation has to be programmed to generate formal functional controls. Accidental methylations would never support local, let alone holistic, homeostatic metabolism.

Neither molecular evolution nor post-biotic evolution can Select *FOR (in pursuit of)*.

3.4.3 The Genetic Selection Principle (GS Principle)

The GS (Genetic Selection) Principle states that biological selection must occur at the nucleotide-sequencing molecular-genetic level of 3'5' phosphor-diester bond formation. Selection must also take place at the level of epigenomic switch settings (e.g., cytosine methylations). After-the-fact differential survival and reproduction of already-living phenotypic organisms (ordinary natural selection) does not explain polynucleotide prescription and epigenetic switching.

All life depends upon literal genetic algorithms. Even epigenetic and "genomic" factors such as regulation by DNA methylation, histone proteins and microRNAs, are ultimately instructed by prior linear digital programming. Biological control requires Selection *FROM AMONG* particular configurable switch-settings to achieve potential function. This occurs largely at the level of nucleotide selection, prior to the realization of

Section 3: How did primordial prescription, and its processing, arise in a prebiotic environment?

any isolated or integrated bio-function. Each Selection *FROM AMONG* a nucleotide corresponds to pushing a quaternary (four-way) switch knob in one of four possible directions. Formal logic gates must be set that only later will determine the folding and binding functions of prescribed proteins through minimum free-energy sinks. These sinks are determined by the primary structure of both the protein itself and by the independently-prescribed sequencing of chaperones. Hexanucleotides also determine Translational Pausing (TP) at the back end of ribosomes.[253,254] TP prescribes protein folding.

Living organisms arise only from computational halting. The fittest living organisms cannot be favored until they are first computed. The GS Principle distinguishes Selection *FROM AMONG existing* function (natural selection) from Selection *FOR potential* function (formal selection at decision nodes, logic gates and configurable switch-settings).

The cybernetic function of the genetic material symbol system *controls* amino acid sequencing and the folding of proteins into molecular machines. The sequencing of nucleotides as physical symbol vehicles is not determined by physicodynamics. It is *dynamically inert* (physicodynamically incoherent; decoupled from physical determinism).[45,59,60] Once instantiated into a Material Sign System, however, this dynamically-inert programming physically determines amino acid sequencing and their dynamic folding space.

The GS Principle attributes the organization of life, along with life's controls and regulation, to *nucleotide sequencing selections and other epigenetic switch-setting selections*. Evolution cannot select at this molecular/genetic level. Evolution cannot program at the decision-node level in pursuit of eventual computationally halting function. Evolution works only on small groups of already-programmed, already-living phenotypic organisms. Evolution is eliminative, not creative, at the programming level. Evolution is nothing more than the differential survival and reproduction of the best already-programmed, already-living phenotypic organisms. The GS

Principle states that evolution cannot select nucleotides or cytosine methylations. No selection pressure at all exists at the molecular/genomic level where programming must occur.

3.5 Molecular evolution offers only two possible kinds of supposed "fitness."

The closest approximations of selection in molecular evolution models are:

1. Greater molecular stability and longevity of certain moieties.

2. Mutual- or Self-catalysis leading to replication dominance (e.g., of RNA analogs)

3.5.1 Greater molecular stability as a means of "selection"

More stable molecules would have decomposed more slowly, and therefore "hung around" (not "survived") longer. "Survival" is a life science term that is totally inappropriate for describing mere molecular stability. An inanimate entity cannot survive because it was never alive in the first place.

Unstable molecules would have broken down quicker, and would have ceased to be present in a prebiotic environment. So, there is a kind of indirect selection going on in this model that merits some serious attention. The proposed mechanism of "selection" is "differential longevity of molecular existence." In other words, some molecules would have lasted longer, which is somehow imagined to correspond to greater "fitness." The macro-evolutionist would argue that we have here a mechanism for "selection" that might explain how evolution got started.

There are some real problems with this kind of back-door, indirect, pseudo "selection." First, mere molecular stability provides no facilitation for any organization. Molecular stability has nothing to do with molecular

Section 3: How did primordial prescription, and its processing, arise in a prebiotic environment?

machine formation and function. The "selection" of differential molecular stability has absolutely nothing to do with life.

Second, many of the most stable molecules in any environment are the most toxic to life. Just because a molecule is more stable doesn't make it useful, favorable or functional to life. Stable molecules would not have contributed much to abiogenesis or ongoing biogenesis of new families of organisms. When stable molecules form (those that are irrelevant or toxic to life), they bind and tie up valuable potential atomic resources needed for molecules that are essential to life. The formation of useless stable molecules consumes "energy available for useful work." Life can ill afford to waste any usable energy.

New more sophisticated prescription and processing require Selection *FOR (in pursuit of)*. Selection *FROM AMONG* could not have programmed life from mere stochastic ensembles or polyadenosine-like homobiopolymers. Selection *FROM AMONG* has absolutely nothing to do with programming life.

The bottom line is that no basis for either kind of "selection" exists in a prebiotic environment. First, no living organisms existed from which to select. Even more important, no basis for selection existed at the decision-node level where each nucleotide had to be selected in proper sequence to prescribe function *before* function existed. Nucleotides needed to be selected to program each triplet block code for each amino acid. In addition, the correct triplet codon needed to be selected to prescribe proper protein folding in the ribozyme. Protein folding is partly prescribed by which redundant codon is used to prescribe a certain amino acid. This in turn controls translational pausing (TP) mechanisms and folding at the back door of the ribosome.[253,254] Use of a different redundant codon would not have worked, even though the correct amino acid was prescribed. Because all known life is cybernetic, controlling choices had to have been made at each decision node—at each configurable switch-setting—at each token selection in nucleotide syntax prior to the existence of any selectable functionality. Any evasion or denial of the need for such bona fide purposeful choices of nucleotides is utterly specious. The absence of Choice Determinism (CD) results only in genetic noise and coding gibberish. Computational success is wholly dependent upon Choice-Contingent Causation and Control (CCCC).[19,29,31,42] Non-trivial computational halting has never once been observed to be produced by chance and/or necessity.

In addition, life often depends upon rapid degradation of regulators like ncRNAs. The effectiveness of most epigenetic switches and their control of gene activity depend upon the rapid degradation of protein enzymes.

The bottom line is that mere molecular stability is not a causal mechanism for protocell formation, and cannot begin to account for abiogenesis. If molecular evolution is based on mere differential molecular stability and longevity, molecular evolution is a non-theory for abiogenesis.

3.5.2 Mutual- or self-replication as a means of "selection"

Auto-catalysis refers to a single RNA-like molecule that catalyzes its own self-replication. Mutual catalysis refers to two different molecules that catalyze each other's replication. Most mutual- or self-replication models involve an RNA analog that simultaneously serves both "informational" and catalytic functions.[255] Here, "informational" actually means nothing more than combinatorial Uncertainty. A stochastic ensemble (random sequence) would have the highest number of bits of Shannon Uncertainty. This has nothing to do with Prescription and its processing of any sophisticated function.

Autocatalytic RNAs are extremely difficult for highly intelligent RNA chemists to purposefully design. They simply do not form spontaneously in any plausible RNA World environmental model. It is extremely unlikely that a mutual- or auto-catalytic RNA analog would have spontaneously formed in the first place. The structural requirements for such sophisticated catalysts is not just complex, but conceptually complex. It is formally ingenious. An artificially engineered self-replicative RNA is a major accomplishment requiring the expertise and ingenuity of the world's finest RNA chemists. Even a single spontaneously self-replicative RNA analog was in all likelihood almost statistically prohibitive in a prebiotic environment.[256-263] About now, we usually hear the inane, "It *could* happen!" The utter *implausibility* of the model is completely and conveniently ignored.[147-153,156] Mere combinatorial complexity, especially with no basis for step-by-step iterative Selection *FOR* potential function, simply does not yield such sophisticated self-replicative molecular machines. (See Section 3.19) Hundreds of such instructional catalysts would have been needed. Their formal organization into a protometabolic scheme also would have been required. An inanimate environment could not have schemed for, or organized, anything.

RNA analogs are included in the model because they are easier to form, and are less likely to break down quickly, than the RNA of current life. RNA analogs do not occur in current life. Many abiogenists just believe that early life would have had to be different because of the many RNA biochemical problems that would have existed in a primordial environment.[264-266] The RNA World was initially the great hope of abiogenesis scientists, until it became painfully apparent how extremely implausible an RNA World would have been.

The first problem with all of these models is that any mutual- or self-replicative RNA analog that might have arisen would have been so highly optimized for its own self-replicative catalytic tertiary structure that it would *not* have been optimized at the same time for any other needed metabolic structure and bio-functions. Because we can artificially produce stochastic ensembles of an RNA analog, or artificially engineer its sequence, does not mean a prebiotic environment could have programmed auto-catalytic prescription into that RNA's sequencing, along with simultaneous protometabolic capabilities in that same self-replicative molecule. Thus, any molecule optimized for auto-catalysis would not be simultaneously optimized for other metabolic functions.

The next problem is that any mutual- or auto-catalytic RNA analog(s) would have massively self-replicated to the point of consuming all physicochemical resources needed for hundreds of other biopolymers and organic bioreactants. Sequence space (Ω) would have been depleted. No other needed moieties could have formed and been selected (not that a prebiotic environment would have had any ability to select any moiety for any reason). Hundreds of other instructive and catalytic biopolymers also would have been required for the spontaneous generation of even protolife. Few other biopolymers could have formed in competition with a self-replicative RNA analog selfishly consuming all biopolymeric resources for itself.

Only one category of selection remains, Selection *FOR (in pursuit of)*. But Selection *FOR (in pursuit of)* requires purposeful programming choices at bona fide decision nodes. Naturalistic science can't possibly entertain *that*! Yet, such selection *must* take place prior to the existence of any living phenotypic organisms.

So what do we have left as a theoretical mechanism to explain the undeniable programming of life? Even primordial genetics, genomics, epigenomics, development, and eventually metamorphosis require programming. So far as selection is concerned, philosophic naturalism has *nothing* with which to work.

An inanimate environment could not have valued or pursued utilitarian goals. No naturalistic scientist could seriously entertain a model of *pursuit* in a prebiotic environment. More specifically, a prebiotic environment did not value, pursue, or seek to preserve a certain binding energy or catalytic function. From a physics and chemistry standpoint, a binding energy is just whatever it happens to be, without preference or relation to formal function.

Minimum Gibbs-free-energy constraints exist thermodynamically. But a primordial environment had no concept of, and no interest in, "usefulness" or "function." The inanimate environment could care less whether anything functions, let alone functions optimally. No perception of foothills or mountain peaks of formal utility exist in a prebiotic environment. An inanimate environment doesn't even perceive foothills of formal function. Whatever happens physicochemically, happens. There's no "good' or "bad" with reference to utility in a prebiotic environment. Inanimate nature is blind to function.

If post-biotic evolution has no goal, certainly molecular evolution had no goal. Thus, Selection *FOR (in pursuit of)* has absolutely nothing to do with molecular evolution.

Humans can declare, after the fact, what binding energies they perceive might have been "helpful" to an organizing metabolism based on prior Descriptive Information (DI), or Epistemological Prescriptive Information (PI_e).[2,21,82] But that injects human epistemology and agent prescription into what was supposed to be a prebiotic model. A subtle and usually hidden anthropocentric role provides the mainspring for every scenario that has been purported to demonstrate a spontaneous naturalistic origin of life. No such spontaneous naturalistic generation of new non-trivial Ontological Prescriptive Information (PI_o), or its processing, has ever been observed or demonstrated.

Well, what about quantum theory? That could program, couldn't it?

Quantum events are statistical. Statistical theory is description, not prescriptive. Programming is prescriptive. All known cells, including the simplest, are highly cybernetic. Even in Mycoplasmas, cybernetics steers events toward pragmatic computational success. Metabolism and life are formally controlled, not just physicochemically constrained. Programming will never be explained by Statistical Theory. Any form of programming is a function of Decision Theory. Life is replete with controls and the most sophisticated regulation known to science. Artificial cybernetics came into existence as a result of the fathers of computer science being inspired by subcellular cybernetics. As if genetics and genomics were not proof enough of formal controls directing life's processes, the ENCODE project[251] has only magnified the importance of life's purposeful configurable-switch settings.

In the case of abiogenesis in an inanimate naturalistic environment, no non-trivial function existed yet for the environment to prefer, not that an inanimate environment would have favored isolated function over non-function. A prebiotic environment would have been utterly blind to function. No motive or ability to pursue isolated functions existed leading up to an eventual protometabolism. Any physicochemical reaction or interaction was just as "good" as any other. No basis, therefore, existed for Selection *FROM AMONG*, in a prebiotic environment.

All the players must be in place at the same time and location for any abiogenesis model to be plausible. This is true whether the protocell model proposes a large number of peptides, protein enzymatic molecular machines, or a much larger number of highly inefficient cooperative ribozyme RNA analogs.[264-267] Peptides and ribozymes are pathetic catalysts compared to protein enzymes. It would take hundreds of highly inefficient ribozymes to begin to organize an effective protometabolism. Their capabilities are just too limited. Yet, just one or two spontaneously-forming self-replicative ribozymes at the same place and time borders on being statistically prohibitive.[243,268]

The 16S RNA subunit has 1540 sequenced nucleotides. The 23S RNA subunit has 2900 specifically sequenced nucleotides. How did all these nucleosides get properly sequenced so as to fold into the correct shapes, or prescribe with code the needed instructions and processing?

In the case of protein enzymes, none exist independent of an already ingenious, conceptually complex ribosome translator/factory that algorithmically processes the conceptually-edited positive mRNA "Turing tapes."[2,21,82] In still another chicken-and-egg paradox, prokaryotic ribosomes themselves consist of a large number of essential proteins in addition to their RNA. In addition, the small subunit depends upon 21 highly functional proteins, each of which has a critically programmed syntax of prescribed amino acids. The large subunit has 31 proteins.[269] How were these proteins formed if they are essential components of the ribosomes that alone can manufacture proteins?

Any naturalistic model that presupposes inanimate nature's pursuit of superior function is totally without empirical and sound experimental support. Strip Materials and Methods of its hidden experimental formal controls, and the experiment produces nothing but worthless tar.

Molecular evolution offers only two mechanisms of pseudo "selection." Neither of these is bona fide "Selection *FOR (in pursuit of)*." Neither mechanism can program new computational function.

3.6 Selection "Pressure"

Environmental selection is often erroneously called "selection pressure." No pressure exists in an inanimate environment to program anything. No such force is exerted by the environment to select for anything. In fact, no pressure exists within living cells to program superior software, or to construct superior firmware.

Ordinary natural selection (NS) is extremely indirect and passive. NS has never been observed to program anything. Natural selection is nothing more than the differential survival and reproduction of the fittest already-programmed, already-living phenotypic organisms. No pressure exists towards improved bio-functional fitness or superior programming. NS does not and cannot operate at the genetic level (The GS Principle).[68,71,83] "Pressure" is a misnomer that arises only out of wish fulfillment. Proponents of "selection pressure" simply wish to imbue purely physicalistic models with innovative formal programming talents that do not exist in a purely physicodynamic metaphysic.

Using the term "pressure," therefore, is a linguistic and rhetorical "sleight of hand." This ploy is used to prop up blind belief in an imagined relentless uphill progress in prescription of sophisticated new functions at the programming level of genomics. Genomics is under no pressure whatever from the inanimate environment in or around each cell. There is not even any pressure from the internal cellular environment to improve subcellular prescription and performance. The cell exercises its preprogrammed "end-user freedom" to optimize its response to environmental stimuli. But the cell's genome is under no pressure to rewrite its own software, or to redesign its firmware. The cell has no such capability anyway.

Selection "pressure" is nothing more than a play on words, with zero scientific substance. No hint of "pressure" exists. An attempt is made to equate secondary "survival of the fittest" with an imagined "necessity" of law to program and optimize formal algorithms. Blind belief is abundantly exercised in a relentless uphill push of nature toward ever more sophisticated programming. Yet, no formal programming choices of any kind are ever exercised by nature. The fittest living species just simply survive better.

Says Edelman and Gally in a PNAS paper, "There is no evidence to support the view that evolution guarantees progress."[270]

What exactly exerted selection "pressure" in a prebiotic environment? What force of inanimate nature pushed physicodynamic and physicochemical interactions towards formal utility? Was it gravity? Was it the strong or weak nuclear force? Was it electromagnetism that pushed molecular collisions towards impressive formal structural organization and integrated metabolic function?

A lot of scientists just fanatically pontificate the doctrine of "emergence" as gospel Truth. "SEZ WHO?" philosophers of science ask. On what scientific basis is this religion being propagated? What exactly is the empirical evidence supporting the dogma of "emergence"? By what natural mechanism would utility have been desired, valued, or pursued in a prebiotic environment?

Natural Selection (NS), therefore, does not and cannot program new organisms into existence (The GS Principle[71]). Selection "pressure" cannot purposefully choose nucleotides at each locus in a DNA positive strand.[155,271,272] "Selection Pressure (SP)" cannot write polynucleotide or polycodon syntax.[54-58,273,274] It cannot program DNA.[2,41,70,71,79,86,275] "SP" cannot prescribe sophisticated new function using a representational symbol system.[43,44] It cannot organize a block code (triplet codon table) designed and engineered to reduce noise in Shannon channels.[7,39,82] "SP" cannot write error-correction codes and parity bits.[276,277] It cannot edit alternative splicing of mRNAs. "SP" cannot anticipate folding needs of all of the polyamino acid polymers, and then prescribe all of the needed chaperone proteins to be separately produced to aid in the folding of each specific future protein that will be needed for metabolic success.[7,278-281]

Can evolution alter existing programming? Absolutely. Does that explain "initial programming"? No. Evolution just presupposes initial programming.

No cell could have come into existence without prior programming. Life is just too conceptually complex and formally organized, even at the simplest level (e.g., *Mycoplasma genitalium*).

Selection *FOR (in pursuit of)* is never involved in molecular OR even post-biotic evolution.

Selection *FROM AMONG* simply favors the small group of the most computationally successful organisms that survive and reproduce the best. With post-biotic evolution, every one of the competing small groups of phenotypic organisms already exists at the time of selection. No "pressure" of any kind exists in Darwinian evolution. Selection is entirely indirect, passive, and only secondary.

In the case of prebiotic molecular evolution, no living organisms exist yet for the environment to secondarily prefer. Zero "pressure" exists to select anything. Something that does not yet exist cannot be preferred.

With both molecular and post-biotic evolution, any "pressure" is an illusion. "Pressure" is vividly imagined in order to make our model "work for us." Never mind whether the model corresponds to objective reality. The only thing this model "works for" is our ill-conceived philosophy, not science. Selection "pressure" simply does not exist.

3.7 "Directed Evolution"

"Directed evolution" is a self-contradiction. If something evolved, it wasn't directed. If it was directed, it didn't evolve. This nonsensical concept of "directed evolution" has no place in scientific literature. Evolution not only has no goal; it is not *steered* toward a goal.

The whole point of evolution theory was to explain "apparent design" without any need of actual design. All efforts to prove that design is only apparent rather than real have failed, however. Every "directed evolution" experiment in the literature has in fact, been directed, just as the name implies, in violation of evolution theory's most fundamental supposition. In addition, evolution cannot possibly explain the programming that would be needed to generate even the appearance of design. It would take more of a miracle for inanimate nature to simulate the degree of apparent design than design itself requires.[282-287]

Why are so many "directed evolution" papers in the literature? The answer is that "directed evolution" is the only way any supposed support for evolution can be generated. Without steering, nothing of any functional significance ever "evolves" into existence. Any random modification of existing PI results only in noise pollution of that PI. Any "editing" by nature results in a self-ordering, law-like tendency that only reduces Shannon combinatorial uncertainty (fewer bits). This in turn only reduces informational and engineering potential.

Contrary to popular belief, an increase in Shannon combinatorial uncertainty (complexity) has nothing to do with an increase in functionality. Maximum uncertainty is randomness. Randomness has never programmed any computational function or engineered any significant new form of functionality. It is true that combinatorial uncertainty is required for choice-contingency and info genesis to be recorded into physicality. A programmer must have freedom from Physicodynamic Determinism (PD) in order to choose from among real options at true decision nodes. Freedom from law-like determinism (PD) generates not only uncertainty, but possibility. We worship "possibility." But possibility never programmed any computational program. Possibility never built a single sky scraper. Only one thing programs computational success and builds sky scrapers: Choice Determinism (CD). CD is the bottom line of "directed." This is why "directed" must be smuggled in through the back door in order to realize the slightest hint of macro-evolutionary experimental progress. We see zero empirical evidence of spontaneous "evolution" producing any new Prescription of non-trivial new functionality. No prediction fulfilments exist of evolution having generated a single new type of organism. Only "directed evolution" produces anything of any worth that is truly new. But if this supposed evolution was "directed," it wasn't evolution at all. It was agent-mediated Choice-Contingent Causation and Control (CCCC).

If anything, the need for "directed evolution" to produce a single new non-trivial function is just all the more evidence against the bankrupt notion of spontaneous macroevolution.

3.8 "Evolutionary Algorithms"

In "evolutionary algorithms," we have a sort of linguistic inversion of the above nonsense phrase, "directed evolution."

An algorithm is a purposeful agent-mediated process of successive steps made in pursuit of some computational or structure-related functional goal. Algorithms consist of a sequence of decision-node programming choices. The generation of "evolutionary algorithms" begins with a pool of "possible solutions" to a "problem." "Problems" and their "solutions" are abstract and formal, not physical. Algorithms require a "fitness function." A "fitness function" is also formal. An agent must be motivated and able to determine when a fitness function is optimized. Choices must be made prior to the realization of optimization. That agent must be able to steer the change between the initial state and the goal state. Inanimate nature cannot and does not deal with problems or their solutions. No algorithm, or its optimization, has ever been generated by anything other than a bona fide "agent" making purposeful choices in pursuit of eventual utility. Inanimate nature, evolution included, does not and cannot generate algorithms. It cannot optimize them, either.

Neither Markov processes nor random number generators have ever been observed to generate functional programs and computational halting apart from hidden experimenter steering. Non-trivial algorithmic optimization simply does not happen on its own. Without painful self-honesty on the part of each scientist, science will collapse under the weight of postmodern subjectivism and solipsism. No evidence exists of evolutionary algorithms occurring spontaneously in a truly "naturalistic" world.

In every peer reviewed paper, "evolutionary algorithms" can be shown from Materials and Methods to example nothing more than "directed evolution." This is usually accomplished through the purposeful choice of which iteration to pursue at each algorithmic step in pursuit of fitness and optimization. Both "evolutionary algorithms" and "directed evolution" are self-contradictory nonsense terms. Both provide examples of nothing more than *artificial* selection, not natural selection.

Life could not have commenced spontaneously in nature without integrated bio-functional *processes* being prescribed and organized into a collaborative effort. We call that collaborative effort, "homeostatic metabolism." Metabolism requires Prescription and Processing (P & P).

That fact would have been true of the very first primordial cell if its "life" were real and sustained. Even a primordial cell would have consisted of many Sustained Functional Systems (SFSs), all complementing each other.

The phrase "Evolution invented . . ." appears all the time in scientific literature. Evolution can't invent anything. Evolution is nothing more than the differential survival and reproduction of the best already-programmed, already-living phenotypic organisms.

3.9 Ontological vs. Epistemological Prescription and Processing (P & P)

Epistemology is the study of how humans know what they think they know. Ontological reality is objective *being*. The ideal of science is for human knowledge to *correspond* to objective being. Human knowledge, even when mistaken, is a part of objective reality. Human beings and their relative, subjective knowledge of ontological *being* are a *subset* of ontological *being*. Ontological reality is bigger than both the *Homo sapiens* species and its knowledge.

If our knowledge fails to correspond to objective being, it is misleading. A tension will eventually develop between what we thought we knew, and what IS.

Humans have a distinct existential tendency to view themselves as the center of the universe. Owing partly to the metaphysical worldviews of philosophic humanism and naturalism, they also tend to see themselves as "the measure of all things." We cannot seem to view reality without getting our own knowledge, understanding and metaphysical beliefs hopelessly intertwined and confused with the way things objectively are outside of our minds. This is known as the human "epistemological problem." We cannot divorce ourselves from our own subjective knowledge, thoughts and perspective. This affects our ideas *about* how life may have begun. It is the nature of the human condition that our minds become an inseparable ingredient of our thought structure.

Nothing puts us in our proper place in the Cosmos quicker than 1) astronomy, and 2) an honest consideration of abiogenesis. Life-origin was in no way dependent upon human mentation. Humans didn't exist yet. Despite our "epistemological problem," science has an obligation to consider what might have obtained in ontological, objective reality prior to human existence, during any presupposed naturalistic emergence of life.

This means that human ideas of uncertainty, information, knowledge, and searches for targets are irrelevant to how life, its programming, processing, and its regulatory cybernetic operation, began. In other words, "Keep human notions of 'information' out of it, stupid." The issue is, "How did ontological, objective Prescription and its Processing (P & P) first get programmed within a cosmos consisting only of chance and necessity, mass and energy?"

A related problem for physicists might be, "How did ontological, objective mathematical laws get written within a cosmos consisting only of chance and necessity, mass and energy?" Cosmologists and cosmogonists have already concluded that the mathematical laws of physics and chemistry only came into existence at the time of the Big Bang. But no one has explained how a random explosion of a cosmic egg could have generated formal mathematical logic. Just as chance and necessity, mass and energy could not have generated the mathematical laws of physics and chemistry, they could not have generated the P & P required by cybernetic life.

The latest cosmos TV series asks the question, "Where did all this information come from? Nobody knows." But the real issue is not just epistemological information *about* the birth of genomics. The real issue is how actual Prescription and its Processing spontaneously arose out of chance and necessity sufficient to organize and bring life into such tightly regulated existence. After-the-fact Selection *FROM AMONG* the fittest already-programmed, already-living organisms cannot begin to address this question.

Abiogenesis researchers must carefully distinguish between previously existing ontological cellular information and their applied mental epistemological constructs of information. Ontological Prescriptive Information (PI_o) cannot be reduced to our finite ideas of it. Our understanding is perspectival, subjective Epistemological Prescriptive Information (PI_e). The more we learn about biological PI_o, the more we realize how much we don't know about it. PI_e pales in comparison to PI_o. PI_o is the standard, not PI_e.

An inanimate environment cannot generate PI_o because a prebiotic environment cannot generate CCCC—it cannot make programming choices or communicate using a semiotic symbol system. Cybernetics and semiotics are formal enterprises, even when they use physical symbol vehicles (tokens). CCCC is a primary formal cause, not a physical physicodynamic effect. The energy requirements for programming and processing a binary decision node are the same, whether a 0 or a 1 is chosen.

Life-origin investigators must stop talking about "information," and start concentrating on ontological P & P of bio-functions and integrative metabolism. What was the source of the Processing that would have been needed simultaneously along with the Prescription, in order for that Prescription to have any value and effect? There will be no escape from the most fundamental principle of science, the F > P Principle (the Formalism > Physicality Principle).[19] (See Section 3.16).

The source of Prescription of sophisticated function, and the ability to process that prescription, is the real issue of life origin."

3.10 Can existing ontological prescription, and its required processing, spontaneously increase in nature?

A tremendous increase in the sophistication of Prescription, and its processing, would have been required for each and every new step up in any macro-evolutionary scenario.

Many layers and dimensions of P & P are currently being uncovered even in the simplest cells. The source of such delicate regulation within any cell is programming. The cell's prior programming prescribes how it will react to most any environmental challenge. A myriad of potential cellular needs had to be anticipated by the cell's genetics/genomics and epigenetic switches. This versatility in programming reveals extraordinary additional layers of sophistication as molecular biological knowledge increases.

Does any geneticist, cell or molecular biologist seriously doubt that "Life is programmed," especially post ENCODE?[250-252] The only hope of making such extraordinary programming seem feasible within naturalistic philosophy is to keep extending the amount of time that chance and necessity could have had to "tinker with," in small increments, programming by "trial and error." But, time is not a cause. Time is especially not Choice Determinism (CD). We also fail to realize that even "tinkering" is still a formal process, a CD process. It is not a Physicodynamically-Determined (PD) interaction, reaction or phase change. Trial-and-error searches are a form of purposeful Selection *FOR* (in pursuit of). "Is it this one, Yes or No?" "Is it that one, yes or no?" Such successive binary queries may constitute a very crude search for desired utility. But these simple binary questions nonetheless constitute a formal search. Inanimate nature conducts no such formal search for utility, let alone potential utility.

Particularly in the Precambrian explosion (the rapid appearance in the fossil record of most phyla), the rate of accumulation of new, more sophisticated Prescription and its Processing (P & P) in biosystems could not have been produced by macroevolution in such few generation times, especially. Questioning only "the rate of information accumulation," however, presupposes that information is accumulating. Is it? Do random mutations increase P & P? Do nonrandom mutations (physicodynamic determinism) cause an increase in P & P? As this entire book shows, neither chance nor necessity, nor any combination of the two, can produce an increase in P & P.

Only one thing can increase P & P: Choice-Contingent Causation and Control (CCCC) also called Choice Determinism (CD). Choice-contingency emanates only from the far side of The Cybernetic Cut. No empirical or rational basis exists for anyone believing in a spontaneous increase in P & P from "natural" causes.

This is not just true with abiogenesis. It is also true in macro-evolutionary models. We can observe slight variation of existing P & P with mutations plus differential survival and reproduction. But we will not see spontaneous increases of prescription of sophisticated new functions arising from the near side of the Cybernetic Cut.[41,42] Chance and necessity, mass and energy, do not and will not act on existing genomes and epigenomes to substantially increase their programming proficiency. It is a logical impossibility for chance and necessity to program at decision nodes and logic gates.

3.11 Chaos theory's self-ordering cannot produce formal organization . . . or non-trivial function.

Self-ordering can occur spontaneously as a purely physical (physicodynamic) event. Prigogine's "Dissipative structures"[288] include phenomena like tornadoes, hurricanes, candle flame shapes, bathtub drain vortices, and sand pile behavior.

The complex interaction of multiple force laws and initial condition constraints produce regularities. Underlying unapparent constraints and law-like influences can shrink probability bounds from a maximally random distribution. This produces spontaneous order. Average distributions become "weighted," or influenced by constraints that reduce combinatorial uncertainty. Maximum order and randomness occur on the same bidirectional vector, at opposite ends of that vector.[6]

Because Physicodynamic Determinism (PD) is hidden in apparent chaos, highly complex causation can result in predictable self-ordered phenomena. This self-ordering can spontaneously appear out of what *looks like* utterly random chaos. But the chaos is only apparent.

Prevailing theories of "chance" often call into question whether anything is truly random. Chance is sometimes viewed as such complex interactions of law-like causation that we cannot clearly unravel the summation and synergistic effects of all the combined causes. Prediction of those interactions becomes nearly impossible. The argument is that they just *seem* random, even though they are actually *caused*.[64]

Tornadoes, however, can almost be predicted to arise from "chaos" given certain complex combinations of initial conditions. Predictability of tornado formation is possible only because of hidden cause-and-effect mediation. A considerable degree of "necessity" is still at play within seemingly random chaos. This interpretation of "chance" just views chance-contingency as a very complex and unpredictable version of ultimate necessity.

While Prigogine's dissipative structures can self-order, they are never organized.[62,81] We cannot equate "order" with "organization." Organization is not synonymous with, and should never be confused with, the natural process of self-ordering. Dissipative structures never accomplish sophisticated functions. Dissipative structures only destroy formal organization; they never create it.

Dissipative structures manifest and generate no Choice Determinism (CD) of any kind. These visual strings of momentarily self-ordered states are purely physicodynamic. They involve no deliberate steering toward utility. No purposeful choices go into the spontaneous self-ordering in nature described by chaos theory. No formal algorithms arise, either. In fact, sequences of "dissipative structures" are most famous for utterly destroying formal organization (e.g., tornadoes, hurricanes), never generating it.

A weather front is not organized. It is only self-ordered. Referring to a weather front as a "weather system" is a misnomer, and is technically incorrect. Sustained Functional Systems (SFS)[30] (See Sections 2.4.4 and 3.11) arise only out of organization and Choice Determinism (CD), not the self-ordering phenomena of inanimate nature. So, a weather front is not a true "system."

What makes "organization" very different from "self-ordering" is that organization is ALWAYS formal, whereas self-ordering is purely physical. Organization can only come into existence through *purposeful choice-contingency,* not by chance-contingency or necessity (law-like determinism). Organization is fundamentally nonphysical, just as all formalisms are nonphysical, such as mathematics, logic theory, language, art, ethics, etc.

Sophisticated efficacious controls don't just happen in inanimate nature. They must be prescribed with programming choices prior to the realization of non-trivial function (e.g., computational success). Programming choices at bona fide decision nodes, in other words, must be made *prior to* the realization of computational success.

And it's not just computer programming that this applies to. No sophisticated function of any kind has EVER been observed to come into existence without purposeful choices steering events and integrating circuits.

But even with the fairly predictable formation of Prigogine's dissipative structures out of "apparent" chaos, we still do not observe any hint of deliberate *steering* of interactions toward pragmatic benefit. The latter occurs only with Choice Contingent Causation and Control (CCCC), never from seeming chance-contingency (See Section 2.7.2).

"Pragmatic benefit" itself is a formal concept, not a primary physicodynamic effect. Inanimate nature is blind to "function." Function is no better than non-function to a non-biological environment. The laws of motion and force laws and constants have no mechanism to identify or pursue utility. Physical factors must be formally organized and controlled with wise Choice Determinism (CD) to achieve organization or sophisticated functionality of any kind.

Purposeful choice-contingency (CCCC) is the most fundamental and essential ingredient of any formalism. The most immediately recognized formalism in science has always been mathematics. Investigators in all fields have always known, and have freely admitted, that mathematics is nonphysical and choice-based. The whole reason it is possible to make mathematical errors is that mathematics is practiced free from law-like determinism. We must voluntarily obey rules, not laws, for mathematical manipulations to be "helpful" to us. Proper use of mathematics is not forced. We can disobey the rules if we want, but only at the expense of formal utility.

Other formalisms include differentiating and categorizing objects, ideas and processes. Categorization is crucial to science. In order to program any form of computation with the simplest binary programming, a 0 must be dichotomized from a 1. 0 and 1 must be categorized distinctly with crystal clear, discrete, excluded-middle, open and shut logic. The logic gate must be open or closed, not stuck half way in between.

Any form of logic theory requires wise choices made according to rules. Symbol systems require purposefully choosing one symbol from an alphabet of symbols. Both sender and receiver must agree on what each symbol and symbol syntax will mean to achieve successful messages. These are all formalisms, not mere physical interactions. All formalisms require discrete choices from among very real possibilities in pursuit of potential

Section 3: How did primordial prescription, and its processing, arise in a prebiotic environment?

usefulness of a product. Even language itself is a nonphysical formalism. Ethics, art, poetry, prose, music and other aspects of esthetics are all formalisms that transcend mere physicodynamics (physical interactions).

We do great disservice to science by failing to properly define and dichotomize "self-ordering" from "formal organization."[62.] Prigogine himself, and many after him, have been greatly confused by sloppy definitions in this regard.[288-290] Some deliberately want to smear the two definitions in order to make their favorite models "work for them." Quality science demands self-honesty, not the propagation of "wish-fulfilment."

The notion of "*self*-organization" is even more ridiculous (See Section 2.10, 3.11, and 3.15). "Self-organization" is a nonsense term when applied to inanimate nature and pre-biotic molecular evolution. The notion is without empirical and prediction-fulfillment support. "Self-organization" is not falsifiable, provides no mechanism, and offers no plausible explanatory power. The concept fails every test of scientific respectability. It is also a logical impossibility for anything to organize itself into existence. The effect of organization cannot be its own cause.

Dissipative structures are not Sustained Functional Systems (SFS)[29,30] (See Section 3.12.5 below). Even the mere self-ordering of a pseudo "system" would quickly dissipate, with no hint of sophisticated function being produced. The self-ordered polyadenosines that spontaneously adsorb onto montmorillonite,[291-293] for example, are potentially informationless. They have almost no Shannon uncertainty, and therefore no Prescriptive Information (PI) retaining capability.

Sustained cell structure, unlike the shape of a candle flame, is no illusion. Many physically-real Sustained Function Systems (SFS)[30] are instantiated into a cell's anatomy and physiology. SFS's collectively contribute to the goal of any cell's being and staying alive through time. The cell's organization will never be found in, or explained by, the self-ordered, momentary "dissipative structure" of chaos theory.

3.11.1 "Dissipative structures" dissipate!

Prigogine called his spontaneously forming structures "dissipative structures" for good reason.[288] "Dissipative" means "momentary," not "sustained." Dissipative structures are anything but stable. The only reason a tornado looks "sustained" is that we are seeing a long succession of momentary dissipative states. This creates the illusion of a stable structure. The shape of a candle flame suggests a stable structure. It is not. Even the continuous string of momentary states that creates the illusion of a stable structure is momentary. It dissipates as rapidly as it self-ordered.

Mere dissipative structures cannot produce bona fide organization, let alone *sustained* organization. Even if they did, they would be gone in an instant, as fast as they formed. Tornado chasers, especially in helicopters, regularly film mile-wide tornadoes that have traveled for a hundred miles, leveling entire towns, suddenly just "disappear into thin air." This is because they are fundamentally "dissipative structures."

Very sophisticated machines and engines are needed to sidestep entropy increase. Such machines and engines are *Sustained* Functional Systems (SFS), not the momentary, highly-ordered dissipative structures of chaos theory.[30,31]

3.11.2 The relation of thermodynamics and statistical mechanics to ontological prescription and its processing

In life-origin science, the temporary and local circumvention of the Second Law is often referred to as "being far from equilibrium." Being far from equilibrium is wholly dependent upon the prescription of formal organization and its processing. Without formal programming, along with the machinery (e.g., chloroplasts; ribosomes) and processing of those programs, equilibrium would be reached very rapidly. Any protocell, if alive at all, would have died long before equilibrium was reached. The local environment may have been "open" to externally-supplied solar energy, for example. But that energy would have remained wasted without formal

Section 3: How did primordial prescription, and its processing, arise in a prebiotic environment?

harnessing, transduction, storage and utilization of that solar energy. Without highly abstract, conceptual schemes, sophisticated molecular machinery, and ingenious processing, solar energy would have only accelerated entropy increase within the local environment. An "open" environment is not sufficient, or synonymous, with an environment that is "far from equilibrium." Without innovative processing, solar energy is just more wasted energy.

> In our opinion, physicodynamic bias only reduces through self-ordering tendencies the vast sequence spaces needed for prebiotic molecular evolution. Even if vast sequence spaces had been available in a theoretical primordial soup, no known mechanism exists for the prebiotic Selection *FROM AMONG prescriptive* sequences. Prescriptive sequences require freedom of selection at each successive decision node. This freedom is antithetical to the self-ordering "necessity" of physicodynamic bias.[62]

The energy immediately following the Big Bang would have been wasted heat energy. No "usable" energy or formal utility is derived from an explosion. Explosions don't even generate order, let along formal organization. Unless there was some formal organizational and cybernetic aspect to what we call "The Big Bang," there is no reason to think that any "*usable* energy" would have been available immediately after the Big Bang. There was plenty of energy, to be sure. But what possibly could have made it *usable* energy? Certainly not the explosion itself. How did this energy get converted to energy "available for work" when "work" is defined intuitively and pragmatically rather than the mere physics definition of work? The mere physics definition of "work" cannot generate cybernetic life. What could have organized and programmed all this mass and energy into such profound and pervasive utility?

We observe both law-like orderliness and formal organization in every direction we look. So we just axiomatically presuppose that this Big Bang Explosion somehow caused both. But why do we blindly believe this?

Certainly not because of science. This leap of faith is purely metaphysical, and totally implausible based on everything we *do* observe. We don't even find a simple wood screw in the forest floor that was formed naturalistically by inanimate nature. How could an explosion have generated the formal, exquisitely rational, mathematical laws of physics, all of the finely tuned force constants (the Anthropic Principle)[318,319], and such highly organized and cybernetic life?

The Second Law does not really describe decreasing order as is so widely supposed. The Second Law actually describes and predicts nature's decrease in organization. Previously-programmed configurable switches are also deteriorating from factors like rust and other forms of oxidation. As the switches fail, the formal choices recorded in those physical devices fail. Decreasing tensile strength of metal alloys leads to airplane structure failures. Deteriorating cement and steel lead to bridge failures. The heat engines that convert steam into usable energy deteriorate. Only the wasted heat energy is left behind to dissipate. Usable energy decreases when the Choice Determinism (CD) mechanisms that alone convert wasted energy into usable energy are progressively lost without new CD innovations being introduced. Usable energy doesn't just happen in inanimate nature. It has to be converted and produced either by a living cell, or by SFSs attributable to life's ingenuity.

When we go to a ghost town in Western United States from the gold rush days, we find highly weathered signs that have lost their painted symbols. The material surfaces onto which the arbitrary abstract symbols were inscribed have deteriorated from wood rot or metal rust. The meaning and function of those signs that was instantiated into physicality from the far side of the Cybernetic Cut are being progressively lost. Although formalisms themselves are nonphysical and immune to the Second Law, the instantiated formalisms (their recordations into physicality) are progressively lost as the physical symbol vehicles deteriorate into simpler and less organized components. The mass/energy of the cosmos remains constant. Sameness, monotony, redundancy, and even order, (not disorder), all increase with entropy increase. Equilibrium is a form of order, where both sides of the partition exhibit the orderliness of sameness, with no uncertainty between sides. What decreases is the organization that alone allowed Maxwell's Demon to dichotomize the two compartments, via choice-contingency, into an energy potential. What opposes equilibrium is not order, but *conceptual* complexity, cybernetic sophistication, integration of circuits, computational success, and the organization of SFSs into functional schemes.

The only instances of sustained exception to the Second law are always associated with life. By life, we are not just referring to the design and engineering feats of human minds. All known life is cybernetic at the subcellular and intercellular levels. The Second Law is clearly circumvented in a sustained fashion at these levels, also. We use the term "sustained" to contrast "dissipative." Life-origin investigators have long appreciated the unique trait of even the simplest life forms to exist far from equilibrium in a sustained fashion.[294-296] Cells, and their many nano-molecular machines, also satisfy the description of a Sustained Functional System (SFS) in that their metabolic schemes are the ultimate in functional systems. Only recently has biology begun to appreciate the extraordinary relevance of "systems biology" in the study of epigenomics though ENCODE.[250-252]

Only Sustained Functional Systems (SFSs) have ever been observed to *forestall,* in a long-term fashion, accumulation of entropy. Progression toward equilibrium and heat death is otherwise relentless. The only solution is a cybernetic one. Cybernetics steers, directs, and programs physical events and objects toward utilitarian success,

Entropy is inevitable without choice-contingent causation and control (CCCC). In the absence of wise Choice Determinism (CD), potentially usable energy dissipates. Prior engineering accomplishments progressively disorganize. Machines and their utility deteriorate.

3.11.3 A cell is a Sustained Functional System (SFS)

A Sustained Functional System (SFS)[19,29,30,42] (See also Section 2.4.4) is any conglomerate of entities that collectively perform some non-trivial function for an extended period of time. Common man-made machines are SFSs. Not only is a living cell an SFS, but many subcellular molecular machines are also SFSs. Take ribosomes, for example. Ribosomes are conglomerates of RNA and many different proteins that collectively perform the sustained function of protein manufacture. Some single-protein molecular machines are so conceptually complex and highly functional through time that they might be considered SFSs. Kinesin, for example, is a motor-protein molecular machine that transports other molecules from one place to another in the cell by literally walking along a connecting microtubule using two legs. We could view the many polymerized amino acid components, the particular sequencing of those amino acids, and the prescribed protein's three-dimensional structure as organized components of the kinesin molecular machine.

When cybernetic formalism is instantiated into physicality, its value is defined in terms of is formal usefulness. All of these formal concepts comprise the essence of metabolism and life. Metaphysical naturalism fails miserably to address this problem.

A Sustained Functional System cannot be organized by chance and necessity alone.[5] A SFS sufficient to generate and maintain an energy potential (e.g., a heat engine), for example, requires trap-door openings and closings that must be controlled by the Choice Determinism (CD) of Maxwell's demon, not merely constrained by Physicodynamic Determinism (PD).[5] The gas molecules in both compartments of this classic thought experiment are inert. Chemistry cannot explain any affinities for, or tendencies toward, utility. Heat agitation in both compartments will always equilibrate through the open doorway unless the trap door is purposefully controlled with the intent of creating a differential energy gradient, or energy "potential."

A worldview limited to chance and necessity cannot explain how any SFS comes into existence. Any attempt to philosophically deny the existence of the third most fundamental category of reality—Selection—makes the generation of any SFS impossible.[21,22,31,42] Not even evolution theory can survive without Selection being real.

3.11.4 A thermodynamically open environment is *not* sufficient to generate a SFS

Apart from formal prescription of function, inanimate molecules will invariably obey the Second Law. Just because an environment is "open" to energy input does not enable it to temporarily and/or locally circumvent the Second Law. In most cases, an external energy source (e.g., the sun's rays hitting a planet or moon) only hastens entropy increase. Mere energy input does not reverse entropy increase.

The earth is "open" to solar energy. That does not make solar energy usable. Solar energy alone does not render earth far from equilibrium. The sun's energy is wasted energy without conceptual mechanisms to harness, transduce, store, or call up that energy when needed. SFSs (e.g., chloroplasts) must be in place and operational to convert and utilize that solar energy.

Machine- and computer-like mechanisms are needed to harness, transduce, store, and call up usable energy when needed. Without these sophisticated mechanisms, entering raw energy only compromises existing organization in that local environment. And, that local environment only dissipates all the more entropy to its periphery.[297,298]

Life is Formally Controlled, Not Just Physicodynamically Constrained. Constraints and controls should never be confused.[61] Constraints consist of initial conditions, boundary conditions, and the law-described orderliness of nature itself. Controls are always formal. Controls steer events toward potential function and bona fide organization (not just low-informational self-ordering).[61,62,81]

The Prescription Principle (See Section 2.8) is not talking about "work" as defined by physics. It addresses the cause of the effect of intuitive, useful, pragmatic work. Useful work is not just a function of an object being moved through space. "Usable" is a formal, abstract concept, not a mere physical displacement.

Materialistic thermodynamicists are often blind to the fact that they have conflated nonphysical formalisms with physicality the minute they start talking about "usable energy." "Utility" and "pragmatism" are formalisms. Formalisms arise only out of Choice-Contingent Causation and Control (CCCC). The most fundamental example of CCCC is the Choice Determinism (CD) exercised by Maxwell's demon in purposefully choosing when to open and close the trap door between compartments.[29] The demon's first purposeful choice of when to open or close the trap door "*in order to . . .*" represents the first incident of prescription. Prescription and its processing (P & P) alone reverses Second Law tendencies.

The demon's purposeful choices and willful actions arise only from the far side of The Cybernetic Cut.[31,41,299] (See section 2.13 and 2.14.) The choice of when to open and close the door constitutes Prescription. The actual opening or shutting of the trap door represents processing—the setting of a physical configurable switch. The actual setting of a physical configurable switch "*in order to achieve some formal goal or purpose*" represents traversing The Cybernetic Cut across the one-way Configurable Switch (CS) Bridge (See Section 2.13 and 2.14). This bridge is the only avenue from the far formal side of the Cybernetic Cut to the near physical side of the ravine. No traffic ever flows in the reverse direction. It is a one-way bridge.

CD alone operates the physical trap door. The physical trap door itself never controls the *decision* of when to change the door's position. Physicality cannot generate traffic flow into the formal domain. Mass/energy does not write the mathematical equations that we call laws.[300-303] Rather, nonphysical, formal, mathematical logic determines, governs and predicts what physical interactions can and will unfold.[19,29,30] This is known as the F > P Principle (The Formalism > Physicality Principle) (See Section 2.12 and 3.16). Formalism traverses the one-way CS Bridge to control all aspects of physicality.

When Maxwell's demon selects a trap door setting "in order to concentrate the faster moving, hotter atoms on one side of the partition," he is Selecting "*FOR*, or in pursuit of, potential function (a usable energy potential)." This is quite different from natural selection, where the environment only "Selects" *FROM AMONG* already-programmed, already-living phenotypic organisms.[21,22,42] Natural selection is always after-the-fact of existence of what is being selected (See Section 3.4 and 3.5). Natural selection (NS) is secondary "Selection *FROM AMONG*." (See section 3.3) NS never Selects *FOR potential* function, especially not at the genetic prescriptive level of symbol system and Hamming block code (triplet codon) use (The GS Principle).[68,71,83] (See Section 3.4.3) Natural selection is nothing more than the differential survival and reproduction of the fittest already-programmed, already-living organisms.

The demon's operation of the trap door, however, represents bona fide cybernetics at the programming level, prior to the realization of any computational halting and formal function.

Formally uninstructed and uncontrolled Physicodynamics invariably progresses towards equilibrium and heat death. All "usable energy available for work" is progressively lost. We need to understand that this concept of equilibrium vs. non-equilibrium has already incorporated non-material factors in violation of a consistently argued metaphysical materialism. "No usable energy left available for work" is not, and cannot be, a purely materialistic concept. No concept, for that matter, is material. Philosophic materialism cannot even argue its own case within naturalistic philosophy or science without violating its own axiom. There is nothing physical about concepts, logic theory, language, debate or the scientific method itself. Representational symbols can be inscribed onto physical tokens, for example.

But, to generate a Material Symbol System (MSS),[44,45,304] those tokens must be purposefully chosen and syntactically arranged. Even material symbol systems remain fundamentally formal, not physical.[82] This is true even though they use physical symbol vehicles (tokens). Symbol systems are as formal as mathematics.

Equilibrium represents the loss of two crucial aspects of reality needed for organization. First, equilibrium represents the loss of all differential dichotomy (e.g., Right compartment vs. Left compartment differences, Open vs. Closed, On vs. Off, Yes vs. No) needed for contingency and Shannon combinatorial Uncertainty (bits of U). Second, equilibrium signifies the loss of all instantiated programming choice-contingency opportunity (CCCC, CD). An agent must have real options from which to choose. All options are the same in equilibrium, meaning that no functional choice potential exists.

CCCC is needed to organize the simplest energy potential needed for a sustained non-trivial heat engine. To establish any usable energy potential, the demon must purposefully choose when to open and close the trap door between right and left compartments containing inert gas molecules at different temperatures and with different velocities. It is only by bona fide *control* of the trap door with Choice Determinism (CD), not by mere physical constraints, that the demon is able to concentrate warmer faster-moving atoms on one side of the partition and colder, slower-moving atoms on the other side. With thermodynamic equilibrium and heat death, Maxwell's demon is left with no choice opportunity—no options from which to choose—no Shannon Uncertainty.

Fanatical materialistic metaphysicians deny the demon the right to *purposefully choose* when to open and close the trap door. Yet, in the absence of purposeful trap door control, no possibility exists of establishing a sustained, non-trivial energy dichotomy between the two compartments, otherwise known as an "energy potential." Notice, however, that not even an energy potential constitutes a heat engine capable of accomplishing sustained, reliable, intuitive, useful work. Additional purposeful choices are required to design and engineer a machine that takes advantage of this energy potential through extended time—a Sustained Functional System (SFS).[30] Only machines (molecular machines included) produce long-term, sophisticated pragmatic benefit. No combination of chance and/or necessity can establish or utilize a sustained useful energy potential and efficient heat pump. An inanimate, prebiotic environment cannot pursue such a goal, either. A simple heat engine might be considered the first Sustained Functional System.

Numerous examples are provided by Metabolism-First life-origin proponents of spontaneous chemical "heat pumps." But these phenomena are never spontaneously organized into SFSs with an integrated purpose, except subconsciously in their minds. Their Freudian wish-fulfillment constructs imaginary integration and functional directionality to their model. Spontaneous "self-organization," and emergence of P & P are just pipe dreams to "make our model work for us" (make our erroneous metaphysical presuppositions *seem* to correspond to the real world).

3.11.5 Entropy is better understood as disorganization rather than disorder.

Thermodynamics equates "randomness" with complete "disorder." It is true that maximum order and randomness are at opposite ends of the same Complexity bidirectional vector.[29] They are both in the same dimension of complexity, only at opposite extremes of the complexity gradient. Randomness is both maximum complexity and maximum disorder. Randomness is also maximum Shannon Uncertainty. Maximum complexity, however, has nothing to do with function or organization. It is nothing less than amazing how few of the world's finest scientists understand this fact.

Maximum order represents the opposite of randomness and the opposite of maximum complexity. But, neither order nor randomness (maximum complexity; maximum Shannon uncertainty) can formally organize anything! Organization is in a completely different dimension from the maximum order vs. randomness/maximum complexity/maximum uncertainty dimension (**See Figure 4, this Section, 3.11.5**).

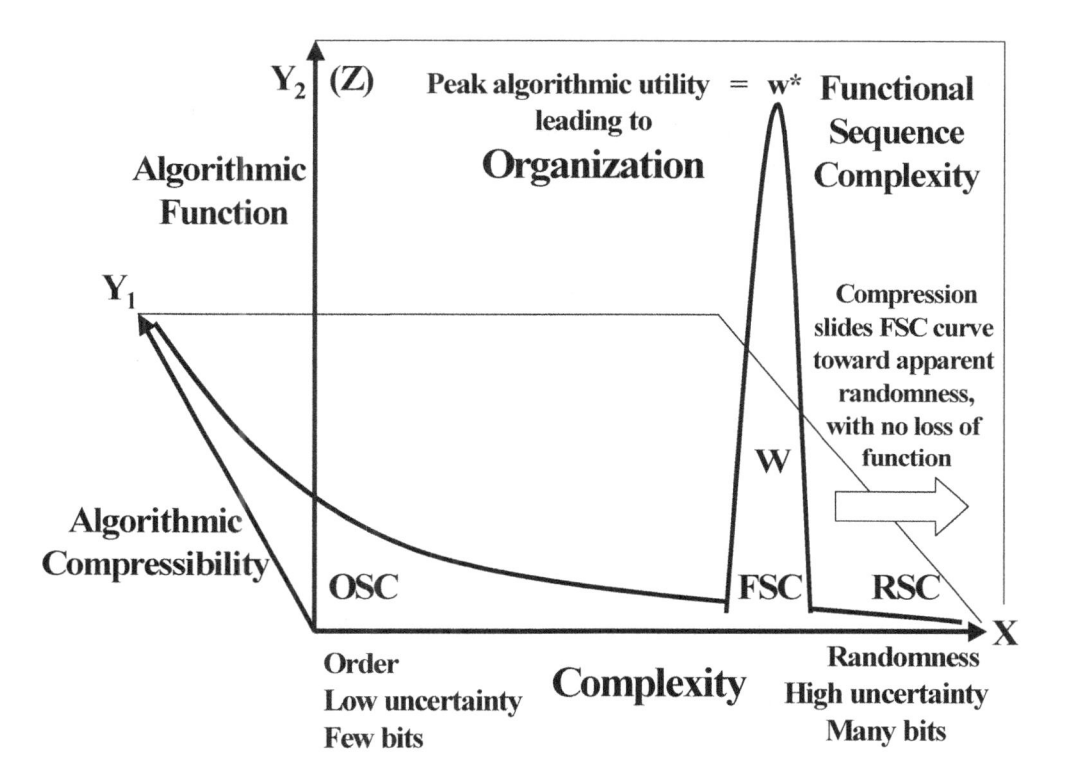

Figure 4. Three dimensional graph showing the relation of *Functional* Sequence Complexity (Y_2 or Z axis) to mere combinatorial sequence complexity (X-axis), and to algorithmic compression (Y_1 axis). The Y_1 axis plane plots the decreasing degree of algorithmic compressibility as complexity increases from order towards randomness. The Y_2 (Z) axis plane shows where, along the same complexity gradient (X-axis) highly instructional and programmed sequences are generally found. The Functional Sequence Complexity (FSC) curve includes all algorithmic sequences that work at all (W). The peak of this curve (w*) represents "what works *best*." The FSC curve is usually quite narrow and is located closer to the random end than to the ordered end of the complexity scale. Compression of an instructive sequence slides the FSC curve towards the right (away from order, towards maximum complexity, maximum Shannon uncertainty, and seeming randomness) *with no loss of function*. Note that both the Y and Z axes can only arise out of Choice Determinism (CD), not out of Physicodynamic Determinism (PD). Nonphysical formalism is required to generate either additional dimension. (Used with permission from reference #6.)

Organization requires Choice Determinism to create an entirely new dimension (Z axis) arising only out of a narrow swath of the complexity dimension gradient (X axis). As physical complexity moves out of this narrow swath of the complexity gradient, either toward randomness to the right, or back toward higher order on the left, organization is always lost. Organization must be free largely from law-like ordering constraints, and yet cannot be produced by randomness either.

Oddly enough, entropy can actually increase with self-ordering crystallization, for example. Crystallization loses potentially useful energy, while simultaneously increasing in order. Crystals still have some energy remaining in their self-ordered structure. They are anything but random. But crystallization can simultaneously move towards increasing order and increasing entropy. Obviously, entropy cannot be synonymous with disorder, as is so widely assumed.[30]

No formal organization or functional information is generated by the increasing order of crystallization. Only high-probability redundancy, unimaginative and boring patterns, loss of uncertainty, and decrease in information-retaining potential result from crystallization.

Organization and Functional Information (FI) cannot be stored in highly-ordered crystal regularities. Organization and FI can only be stored in crystal *irregularities*. Only the irregularities possess the Shannon uncertainty (contingency) needed for FI retention to be possible. But chance-contingency has to be formally converted to choice-contingency for FI retention to be possible.

Functionally, entropy is probably best thought of as maximum disorganization rather than maximum disorder. *Increasing entropy results in increasing loss of organization.* Conversely, anything that increases organization is usually found to be locally and temporarily decreasing entropy. Life represents the ultimate in increased organization from an otherwise self-ordered or random equilibrium state. What keeps life alive is its formal organization and ability to pursue and accomplish functional, computational, integrated-circuit metabolism far from equilibrium. Take away formal programming, and any cell will promptly die.

What so few understand is that only formalism, not physicodynamic determinism, is capable of generating true organization and sophisticated function. All formalisms are choice-contingent. Chance and necessity cannot organize, program or instruct. Chance and necessity cannot pursue non-trivial functionality. Chance and necessity, therefore, cannot possibly decrease entropy. The Second Law governs chance and necessity under all circumstances, barring the intervention of formal Choice Determinism (CD).

There is no excuse for the scientific community to continue confusing "order" with "organization." Organization is always formal and choice-contingent. Self-ordering is not the same as formal organization; disorder and disorganization are not synonymous. Any formal organization already instantiated into physical medium always progresses secondarily toward disorganization as that physical medium approaches equilibrium. As dichotomy and categorization are lost, uncertainty and complexity increase. Organization and functionality decrease as any system moves toward equilibrium and randomness. The system becomes progressively less able to generate and store usable energy to do formal, pragmatic work. It becomes more disorganized. Notice that both sides of the partition being equal (equilibrium) are a form of order, or sameness, not organization.

Any sustained reduction in entropy requires purposeful selection from among real options in pursuit of utility and organization. Chance-contingency is not sufficient. No law of physics can circumvent the Second Law, either. Choice-contingency is required. In addition, formal functionality always progresses towards dysfunction as the Second Law tendency toward disorganization unfolds. This is not because the Second Law affects Prescriptive Information (PI) or Prescription and its Processing (P & P). PI and P & P are both nonphysical and formal. Neither can be affected by the Second Law. What is affected is the physical medium into which PI or P & P have been instantiated, or are processed. That physical medium *is* subject to the Second Law. As this physical medium progresses toward maximum randomness, the retention of instantiated PI, and the ability to prescribe and process in that medium, is progressively lost.

Scientists who should know better confuse this by thinking that the Second Law causes the deterioration of PI. It does not. Usually, their confusion stems back to their starting metaphysical presupposition of physicalism. Once we start our reasoning based on the axiom of materialism / physicalism / naturalism, then prescriptive information *has* to be physical. Because philosophic naturalists fail to realize that their starting axiom is faulty, their reasoning fails to correspond with objective reality.

Once again, PI and P & P are formal, not physical. Only their *instantiation into a physical medium* of retention, transmission or processing deteriorates, according to the Second Law, not the PI or P & P themselves. Physical laws have no effects on formalisms.

3.12 Constraints vs. Controls

Life is precisely controlled and regulated, not just constrained.[61] All known life is cybernetic.[19,30] Cybernetics is the study and application of control mechanisms.[41,42,299] Cybernetics is a function of Decision Theory, not Stochastic Theory.

The "buzz word" of contemporary molecular biology today is "regulation." Life is formally controlled, not just physically constrained. Life is *programmed,* genomically and epigenomically. Many integrative controls are de-centralized and dispersed throughout the cell. Formal prescription of bio-function is instantiated into material symbol systems, configurable switch-settings, integrated circuits, sense and antisense linear digital programming strings. Along with algorithmic processing, these formalisms prescribe, organize, steer, compute, regulate and manage life.

Neither naturalistic constraints nor fixed, forced law-like "behavior" can produce computational programming. We put "behavior" in quotes here because Necessity technically precludes behavior. Behavior has the connotation of free-will choice-contingency. Forced behavior is not really behavior. We are simply observing the effects of physical causation. Good, bad, or errant behavior presupposes freedom from law-like determinism. We throw people in prison because of bad behavior specifically because they could have chosen otherwise. If their behavior were mere effects of deterministic causation, punishment would be altogether inappropriate. Steering a ship with a rudder is a form of true *behavior*. That steering is not constrained. It is controlled. The same is true of programming decision nodes.

The birth of functional controls cannot be explained or derived from mere physicodynamic constraints. The physical laws and resulting constraints are *pragmatically indifferent*. Materialistic cause-and-effect chains under no circumstances provide instructions, directions, steering, programming or prescription of computational success.

Constraints are not controls until they are chosen and selectively used by agents to achieve control. Inanimate nature has no ability to marshal its constraints into cybernetic or engineering function. Constraints just happen, with no agent effort. Nature has no intent or competence to steer cause-and effect chains toward practical utility. When we read "controls" into natural process "constraints," we do so illegitimately. No true controls exist in inanimate nature.

3.13 Laws vs. Rules

For the moment, to understand what "programming" and "prescription" mean, we need to carefully discern between *rules and laws*. *Rules* govern voluntary and arbitrary choice-contingent behavior. The *laws of physics* militate cause-and-effect determinism (necessity).

Laws constrain; they do not control. To control is to steer. Where there is no freedom of choice, steering is not possible. Laws describe an orderliness that forces outcomes. This is the very reason we are able to predict outcomes in physics. *Laws produce order, not organization.* No law of physics forces a trustworthy computation either. Organization and computational success are formal and choice-based. Little flexibility, other than heat agitation and the complexity of interacting causes, exists to produce chance-contingency in inanimate nature. But such contingency never generates choice with intent, formal computational success, engineering prowess, or true organization. *The laws and constraints of inanimate nature operate without regard to pragmatic goals.*[40,154,156,305] To look to laws (especially to "yet-to-be discovered," imagined laws) as an explanation for the derivation of formal controls of physicality is not only empirically unfounded, it is logically fallacious (a category error). No fixed high-probability law can produce algorithmic organization.

Laws are not rules. Rules are not laws. The formal laws of physics constrain the orderliness of physicality. In that sense, physicality is algorithmically compressible into parsimonious laws. We call such laws "elegant" because of their ability to reduce reams of seemingly pattern-less data into a concise description and prediction of motion and physical interaction.

Mathematics is governed not by laws, but by rules. If mathematical manipulations were governed by laws, there would be no possibility of mathematical errors. Our behavior, when manipulating mathematical expressions, must voluntarily obey rules if our calculations are to be helpful. When we perform measurements and mathematical manipulations, we are actually programming. We are making behavior choices according to rules. If we choose to obey the rules, our measurements and calculations are accurate. Our computations produce pragmatic benefits. If our choices disobey the rules, we experience a great deal of pragmatic disappointment. The delusion of having a trustworthy computation that is, in reality, untrustworthy, is worse than no computation at all.

Even the physical laws, however, are actually mathematical formalisms. Physicality is ultimately *governed* by nonphysical formalisms (the F > P Principle, the Formalism > Physicality Principle).[19] (Section 3.16) These formalisms and their sets of rules comprise the real laws of cosmic existence and behavior, not what we normally think of as the "physical laws." Laws, once instituted, preclude freedom of choice. Necessity makes programming of any formal, pragmatic, non-trivial functional success impossible. Neither chance nor necessity can program in any sense.

In a much more fundamental sense, that we will not be able to delve into here, the rules of formalisms that govern mathematics and logic theory, for example, actually arise out of the equivalent of unbreakable fundamental prescriptions that could be called laws. In the same sense, the mathematical equations of physics are called "laws." An example is how we call the formal rules of logical inference "the laws of logic." The "laws" of logic seem as immutable as the physics equation $e = mc^2$. It makes no sense, cosmogonally, to believe that randomness generated the mathematical laws of physics in the Big Bang. Mathematical logic is ultimate logic. Logic, and the laws of physics, could not have been generated from the explosion itself. From the only sensible pre-Big Bang perspective, all mathematical physical laws, and the rules of formalisms, could only have originated out of arbitrarily-chosen logical, rational directives.

To us, after the fact of their issuance, the laws of physics all seem like "necessity." In reality, they were all arbitrarily chosen, and are manifestations of ultimate formalism, not physicality. This is the essence of the Formalism > Physicality Principle (The F > P Principle)[19] (See section 3.16) . Not only the rules of formalisms, but ultimately the laws of physics, were arbitrarily chosen and instituted in the Big Bang "creation" event. But, from our finite perspective from within present-tense physicality, the orderliness described by laws appears to be "necessary." Laws must be viewed differently from "arbitrarily-chosen" rules from a practical, everyday standpoint.

Formal *rules* (not physical constraints) are required to create a symbol or sign system. (Sign systems typically employ pictures, rather than abstract symbols.) Both sign and symbol systems are formally representational. These rules must be voluntarily followed to regain Prescriptive Information (PI) from its physical instantiation.

Rules can be ignored or obeyed. Constraints and laws are "forced," unless artificially chosen, manipulated, and implemented by experimenters. *Choices* of constraints can serve as controls.

Choices *of* constraints can also be governed by rules. But in natural-process physicodynamics, constraints have never been observed to engineer non-trivial computational function. The inanimate environment has no design and engineering prowess.

Laws are rate-dependent. Rules are rate-independent.[306, pg. 195] Their descriptive modes are incompatible. Laws and rules are irreducible to each other. Symbols and symbol systems are time and rate-independent.[306, pg. 195] Physical laws are always functions of rates. They are derivatives of variables with respect to time. We are not talking here about the ordering of events or the various intervals of time. The concept of time itself is irreducible in physics. Symbol systems can do their representing now or later, quickly or slowly.

Time is not arbitrary or contingent. It could not be other than what it is, though different scalar factors can be used to describe it. Time, like space, energy, and momentum, are axiomatic or "properly basic" to physics. These basic concepts, argues Pattee, form the very language of physics. They are irreducible starting concepts.

Rules, unlike laws, have no rate dependency. They cannot be expressed as functions of a derivative with respect to physical or time scales.[306, pg. 195]

> All linguistic operations and all computations, in so far as they are defined by rules, cannot be functions of the rate of writing, reading or

computing. That is, what you mean by a statement of policy or the value of a calculation cannot depend on how fast you speak or calculate.[306, pg. 195]

The fact that rules are rate-independent places them in a category outside the mathematical systems of physics. Rules are not subject to the constraints and laws of physics until after those rules are instantiated into physicality. Even if our discussion were limited to Pattee's epistemic cut, rules cannot be subjected to the reign of the alleged "queen of the sciences." Says Rosen,

"Rules are not interpreted in the same language as laws."[306, pg. 195]

It is often counter-argued that logical rules can be executed by physical switching networks, and that switching networks obey the laws of physics. But it is a *non sequitur* to argue that the rules of logic are predictable from the laws of physics. The rules of logic are merely being instantiated into a physical medium for memory storage and for transmission. But the rules are not reducible to the physicality into which they are instantiated. There is a one-way flow of P & P from the far side of The Cybernetic Cut into physicality. One cannot go backwards from the physical state to recover the rules of symbol vehicle selection that established that selection. This is true at both the individual decision node level of symbol selection, and it is true of the overall programming level of integrated symbol sequence selection.

The rise of P & P had to have occurred early in the presupposed evolutionary history of life. Biopolymeric messenger molecules were instructing bio-function not only long before Homo sapiens existed, but also long before metazoans existed. Many eubacteria and archaea depend upon nearly 3,000 highly coordinated genes. Although genes are edited and alternatively spliced, they are still linear, digital, cybernetic sequences. They are meaningful, pragmatic, physically instantiated recipes.[6]

For prescriptive information to be conveyed, the recipient must understand what the source meant in order to know what to do with the signal. It is only at that point that a Shannon signal becomes a bona fide message. Only shared meaning is 'communication'.[6]

3.14 The Universal Determinism Dichotomy (UDD)

Two kinds of Determinism exist:

1. Physicodynamic Determinism (PD)

2. Choice Determinism (CD)

3.14.1 Physicodynamic Determinism (PD)

Physicodynamic Determinism (PD) is the classic cause-and-effect determinism of naturalistic physics and chemistry. Cybernetics and other formalisms that arise from Decision theory rather than Stochastic Theory never violate PD. But PD cannot possibly explain CD.

3.14.2 Choice Determinism (CD)

The essence of programming is Choice Determinism. Successful computation is *determined* by none other than purposeful programming choices. Halting and successful computation are the *effects* of the *cause* known as Choice Determinism (CD).

In the absence of Choice-Contingent Causation and Control (CCCC) (also termed in the literature, Choice Determinism [CD], as opposed to Physicodynamic Determinism [PD][physical law]), variation is nothing more than noise. Noise can't program anything, let alone computational success, computer architecture, or highly integrative and interdependent metabolic schemes.

Notice that AFTER programming choices are made, no Shannon uncertainty remains. Shannon uncertainty has a negative sign in front of it. Choice Determinism is positive and theoretically absolute. Choice Determinism alone is not certain to produce pragmatic benefit, however. One needs empirical observation of computational success (halting) to be certain that all of the programming choices were wise (functionally optimized) ("the halting problem").[12]

Programming is a function of Decision Theory, not Stochastic Theory. Stochastic theory is descriptive. Decision theory is determinative. Decision Theory possesses the capability of *prescribing* organization and formal function. Mere variation does not. Chance and necessity are the only two categories of reality ruling non-teleonomic "variation." Neither coin flips nor physical law can program cybernetic or semiotic systems. Life is undeniably dependent upon both cybernetics and semiotics. "Teleonomic" is, of course, nothing more than "teleological." Philosophic naturalism just renamed it, hoping no one would notice. There can be no substantive difference between the two as long as Selection *FOR* (in pursuit of) is required.

In inorganic and organic chemistry labs, students get graded on how well they measure out their reactants, and how precisely they follow instructions. The amount and purity of their lab product reflects how faithfully they process the lab instructions (recipe). Chemistry isn't blamed for a poor yield. The chemistry is the same for all students. The student is blamed, and graded accordingly. Why? The answer is that the student is responsible for purposeful choices, and for following the formal rules and instructions that optimize purity and yield of the product.

Most programming improvements attributed to microevolution are in fact switched on/switched off controls of existing PI. The programing was already there, just inactive until switched on. Bacterial resistance to antibiotics is a classic example. Under no circumstances could bacteria, even with a generation time of 20 minutes, evolve new prescriptions of function (resistances to so many different antibiotics) in just 2-3 days of exposure. That's not even sound evolution theory. The PI needed to adjust to antibiotic challenges was already there, in dormancy, waiting for environmental challenge and the need to be activated. The extent of pre-programmed sophistication is just all the more impressive.

3.15 Complexity does not equal abstract concept, organization or functionality

"Emergence" is the term used in an effort to give substance to belief in self-organization. Emergence describes a supposedly spontaneous rise in integrative utilitarian controls and formal functionality of a system. The contention is, "Organized bio-function didn't have to be prescribed. It just emerged," meaning, "it just happened."

The first topic of discussion should be to ask HOW, by what mechanism, did organized bio-function just "emerge?" What is the scientific basis for this notion? How was it established that faith in "emergence" is anything more than blind-belief and superstition? Are we sure that the notion of emergence is not just a product of having to remain logically consistent with a purely metaphysical, materialistic, philosophic commitment?

Relentless research into mere "complexity" in an effort to elucidate the key to life's controls has led nowhere.[313] Mere complexity cannot steer physicodynamic events toward formal organization, biochemical productivity, energy capture, transduction, storage and utilization.[6,31,62,70,85,86,275,314] Maximum complexity is randomness, not formal function. Local and temporary movement away from equilibrium arises only from expedient operation of Maxwell's demon's trap door.[29,30] Life does not spring from complexity. Life springs from quality parts production, imaginative integration of those components, efficient management, constant negative and positive feedback regulation, and quality control. In short, cellular organization and integration of its component systems into a holistic metabolism require expert programming.[26,27,315]

Life is hardware, wetware and software. Programming is a form of Prescriptive Information (PI).[2,6,7,20,39]

Regulation, the new "buzz word" of contemporary systems biology, is a form of control that must be formally prescribed, not physically constrained.

Sinewave patterns prescribe no specific function. Pattern is not synonymous with prescription. Pulsar signals are not meaningful messages. Wave-induced redundant patterns in the bay-shore sand serve no unique prescriptive function. They simply reflect redundant, unimaginative physicodynamic interaction, such as the moon's gravitational force on bodies of water.

Complexity has little to do with PI. Much confusion results from not leaving behind pervasive faulty concepts of "information," with all of their illegitimate epistemological definitions and measurements, and concentrating on the ontological Prescription Principle.

3.16 The Formalism > Physicality Principle (F > P Principle)

The most fundamental principle of science is The F > P Principle, even more fundamental than the Laws of Thermodynamics. The Formalism > Physicality Principle states that Formalism not only describes, but preceded, prescribed, organized, and continues to control, regulate, govern and predict physicodynamic reality and its interactions. The F > P Principle is an axiom that defines the ontological primacy of formalism. Formalism is the ultimate source of all aspects of reality, both nonphysical and physical. Formalism organized physicality before the fact of physicality's existence. Formalism gave rise to the equations, structure and orderliness of nature.

The F > P Principle further states that fundamentally reality's organization is arbitrarily chosen—rule and choice-contingency based, not indiscriminately forced by an infinite regress of indifferent cause-and-effect determinism. This includes the Choice Determinism (CD) that determined what the mathematical laws and law-like order would be. Even the laws of physics, in other words, are choice-determined. Big Bang cosmology affirms that physical orderliness and physical interactions did not exist from infinity, but began within finite time.[316,317]

The Prescription Principle (Section 2.8) is closely aligned with The Formalism > Physicality Principle.[19] The Prescription Principle states that organization can only be prescribed by formal purposeful choices. It cannot arise spontaneously out of chance and necessity, mass and energy. As we saw in Section 3.11.5, order can arise out of chaos, but not formalisms such as "organization." Only purely metaphysical imperatives, not science, force us into accepting the dogma that inanimate nature can generate formalisms.

The F > P Principle alone explains why the scientific method must be conducted in a rational manner, why the applicability of mathematics to physical interactions is reasonable rather than unreasonable, and why formalism can reliably predict physical interactions. Only formalisms can measure, steer, manage and predict physicality. Physicodynamics merely constrains; formalism controls. Only formalism could have organized the exquisitely fine-tuned force constants (Anthropic Principle).[318,319] The F > P Principle alone explains why the applicability of mathematics to physical interactions is reasonable rather than unreasonable,[300-303] and why formalism can reliably predict physical interactions. The quest for a mathematical unified field of knowledge presupposes the F > P Principle.

Physicality cannot even spawn a study of itself—physics—because physics is a formal, nonphysical enterprise. Nothing within the "chance and necessity" metaphysical belief system of materialism is capable of generating language, formal logic, computation, mathematical relationships and cybernetic control.[320]

Philosophic naturalism seeks to deny the F > P Principle. Yet, even before science is discussed, the worldview of naturalism is found to be a purely mental construct. Naturalism is a metaphysical perspective—an abstract, conceptual philosophy. There is nothing physical about "naturalism," or any other "-ism," scientism included! Science itself is a bundle of formalisms. There is nothing physical about "science." For a starter, science is an epistemological system. Every aspect of scientific method involves formalisms such as language, logic theory, mathematics, representationalism with measurements, mathematical symbols, equations and their manipulations. Specifically, the study of inanimate physicality, physics, is also governed by formalisms.

Consider the role that mathematics plays in physics. Almost every physical law is a mathematical equation or logical statement. There is nothing physical about logic theory and mathematical equalities and inequalities. Science depends upon computation at every level. There is nothing physical about computation. This means that naturalistic science is not naturalistic! "Naturalistic science" is a self-contradiction. Formalism is undeniable, and inescapable. It cannot be reduced to mass and energy.

Logical consistency is a requirement of logic theory. But the logic theory inherent in mathematics, language, and cybernetics is formal, not physical. If the worldview of materialism/physicalism/naturalism accurately corresponded to the way reality actually is, formalisms would not exist. Physicalism maintains that everything that exists is physical. To defend the idea of physicalism, one must resort to nonphysical logic theory. The more one argues in defense of physicalism, the deeper a hole one digs for oneself. A logical inconsistency exists in having to employ nonphysical formalisms to support the contention that physicality is all that exists. Even if one succeeded, he or she would only return to physics, the study of physicality, which is 70% formal mathematics. Anything in physics that is not mathematical requires language to convey, which is also formal rather than physical.

Section 3: How did primordial prescription, and its processing, arise in a prebiotic environment?

Empirical data requires sensory data. Neurophysiology has not been very successful at reducing "empiricism" to mass and energy. If sensation is to have any value, sensation requires mental interpretation and willful response. Empiricism plays a major role in science. Yet, empiricism is formally mediated, and formally acted upon.

Experimental design requires devising sophisticated formal schemes such as double-blind studies to eliminate prejudice and test the objective reliability of our observations. We demand adherence to certain ethical standards. We employ statistical analyses. We formally categorize and organize results into logical inferences. We logically deduce conclusions from initial unproven axioms. We abduce: infer from the best model or explanation. We induce from varied data to generalized "laws." All of these activities are formal. Euclidean vs. non-Euclidean geometry involves different axiom sets. Even the study of physicality itself, physics, is nonphysical and formal.

Materialism, physicalism, and naturalism are all self-contradictions. Any "–ism" transcends mass/energy interactions. The very existence of an "–ism" disproves physicalism's most fundamental contention. "Materialism / physicalism / naturalism" is dead. No "-ism" is physical, including naturalism.

The laws of physics could not manifest the predictive powers over physicality that we observe unless formalism controlled physicality. The laws of physics do not just describe physicality, they govern physicality. We are not talking about the equations themselves here. We are talking about the arbitrary control over physicality that the prescription of these equations represents, and originally established immediately before the Big Bang and Planck time began. Plank time is the smallest time measurement possible, 10^{-43} seconds. This is the only reason we are able to refer to the law-like behavior described by the mathematical laws of motion as physical "necessity." Physicality is governed by nonphysical formalisms (The F > P Principle).[19] (See Section 3.16)

Still more evidence of the reality of formalism's dominion is that we have to plug into the equations known as "laws" formal *representations* of initial conditions, not the initial conditions themselves. As physicist Howard Pattee has pointed out many times, the measurements themselves are also formal rather than physical. The genotype-phenotype dichotomy also constitutes an epistemic cut.[16,321]

Von Neumann[322] was the first to appreciate the problem of "complementarity" as it relates to the subject/object dichotomy. It was further expounded upon by Pattee[13,323-326] and later by Hoffmeyer under the name of "code duality:"[321]

> Complementarity is an epistemological principle derived from the subject-object or observer-system dichotomy, where each side requires a separate mode of description that is formally incompatible with and irreducible to the other, and where one mode of description alone does not provide comprehensive explanatory power.[323, pg. 191]

But the problem is far more perplexing than a mere epistemic cut. The problem is not limited to knowledge, or lack thereof. The real problem is one of *causation of initial formal prescription* in a reality that naturalism metaphysically presupposes is limited to chance and necessity. Neither chance nor necessity, nor any combination of the two, can prescribe or program anything. Formal relationships *control* physical events, not just describe physical events. Thus, not only does an epistemic cut[9,12,13,327,328, pg. 36,329] exist regarding human knowledge of laws and measurements of initial conditions, but a *Cybernetic Cut* also exists[41,42] (See Section 2.13 and 2.14). Physicality itself cannot explain being subject to formal mathematical rules. Mathematical representations of initial conditions are required to use laws. The laws of physics are fundamentally non-physical and formal.

In addition, this Cybernetic Cut was in operation prior to the existence of *Homo sapiens*. It is not a function therefore, of human epistemology (See sections 3.9 and 3.10). Ontological P & P always flows from the formal world into the physical world across the one-way CS Bridge (Section 2.14).

3.17 No targets existed in a prebiotic environment, and no searches were conducted

Even some who understand life's absolute need for Prescriptive Information (PI) seem to limit their research into PI by thinking primarily in terms of "searches." But an inanimate, prebiotic environment did not conduct, and could not have conducted, any searches for anything, let alone for formal utility! Formal utility did not exist in a prebiotic environment to be found by any search, even if one had been conducted. Our statistical models of information genesis are hopelessly corrupted with 1) incorrect definitions of information, and 2) investigator involvement inherent in the search models themselves.

Trial-and-error searches are nonetheless still purposeful searches. They may be crude, inefficient searches. But they still represent purposeful pursuits. Purposeful pursuits are teleological. No prebiotic environment could have conducted a "trial-and-error" search.

Discussions of "searches for targets," therefore, are irrelevant to life-origin studies. They reflect scientists' mistaken orientation around their own motivation and epistemological notions of information rather than the non-subjective, ontological P & P required to generate even the simplest life.

In the end, it is not really about searches, targets, goals or even information. Nothing existed in a prebiotic environment to be able to utilize information. Processing requires as much, if not more, Prescriptive Information (PI) than the programing itself.

Searches require motivation. Searches are conducted in pursuit of utility. Searches presuppose value judgments. Utility must be valued to foster a search for any target, and to define the type of search which is most effective. A prebiotic environment valued nothing over anything else. A prebiotic environment also manifested zero motivation. Motivation and valuation are formalisms, not physicodynamic states or interactions. Chance and necessity cannot produce formalisms,[6,19,21,22,29-32,39,42,62,81,82,84] valuation and motivation included.

An inanimate environment contained no "targets." Targets are mental constructions of agents. Granted, a target could be an actual physical location. But, for such a location to become a "target" requires a mental construction that predefines that location as "the *desired* location." No such mental constructions existed in a prebiotic environment, in the first forming protocells, or even in the first primitive cells. No location was any better than any other location. But the target of any life-origin scenario would have been almost infinitely more *conceptually* complex (not just combinatorially complex) than any ordinary target location. It would have been a monstrous constellation of incredibly integrated computational, cybernetic targets, all feeding into the meta-target of "holistic metabolism." Not even the latter does justice to defining even the simplest known forms of "life."

Too many discussions of "information" only secondarily describe it, or search for it, once it exists. It's not about searches. It's about a priori *prescription* of creative new formal function, something that an inanimate environment cannot do. A prebiotic environment could not have prescribed or processed future formal function. Life origin is ultimately more about the origin of cybernetics (control) than biochemistry. Discussing mere Descriptive Information (DI) does not address the problem of ontological P & P genesis or abiogenesis. DI simply presupposes previously existing biological instructions with no explanation whatever for their derivation through natural process. Similarly, "searching for" something that already exists in a haystack also presupposes the prior existence of the object of that search. Biological PI must pre-exist to be "found" in any search, and it also must pre-exist life itself for life to come into existence. How did instructions, and the processing of those instructions, arise in a prebiotic environment of nothing but mass and energy?

Section 3: How did primordial prescription, and its processing, arise in a prebiotic environment?

What in any existing genome would make that genome want to "search" for the "target" of being a higher organism? No such search or target exists in the genome of living organisms. The fittest of each species simply survive the best. That is all that is involved in evolution—differential survival and reproduction of the fittest already programmed, already-living phenotypic organisms. Evolution does not search for targets, or have any goals. The problem of macroevolution is no different from the problem of abiogenesis in the final analysis.

If "evolution has no goal," certainly molecular evolution has "no goal." Yet without goal and purposeful pursuit of that goal, the unlikeliness of even the simplest ten-step biochemical pathway is statistically prohibitive by hundreds of orders of magnitude. The notion of the "spontaneous generation" of life is definitively falsified by the Universal Plausibility Principle of general science.[80,330] The original First Law of Biology (All life must come from previously existing life) stands just as unfalsified today as it was when first proposed.

Even so-called "trial-and-error" notions constitute *a search* with an inherent goal. Evolution has no goal, not that Darwinian evolution existed in a prebiotic environment.

Naturalism has no answer to the question, "What was the initial source of P & P when it comes to abiogenesis or macroevolution theory?" Naturalism also has no answer to the question, "How did P & P keep spontaneously increasing as each new more sophisticated genus supposedly macro-evolved?"

Life origin is about *Prescription* of sophisticated future function---function that does not yet exist for the environment to prefer at the time that prescription must be conceived and processed.

Thus, the first step in elucidating any naturalistic life-origin scenario must be to explain the origin of ontological (objective, independent of human knowledge) Prescription and its Processing (P & P) at the subcellular level. Not even a protocell could have "self-organized" without the direction and regulatory controls of PI_o. There was absolutely no basis for Selection *FROM AMONG* existing function, let alone potential function. The prebiotic environment could have cared less whether anything functioned. It would not have preferred fitness over non-fitness.

To pursue discussions of searches for targets accomplishes nothing. It merely presupposes the motivation and valuation of scientists whose mental constructions did not exist in an inanimate environment. It is no easier to explain "a search for a target" in a prebiotic environment than to explain the birth of formal instructions, halting computation, organization of component parts into all-or-nothing holistic functional machines. All of the latter are formal, and could not have existed in prebiotic nature.

3.18 Intuitive work vs. the physics definition of work.

What is work? "Employment, a job, a vocation or occupation" are all one kind of work that isn't relevant to our discussion here. Other synonyms for "work" might be, "labor, effort, exertion, toil, or even drudgery." But they are not very relevant, either, to what goes on within any cell and organism that we might want to call, "work."

Next is the physics definition of work. Naturalistic scientists will quickly gravitate to this definition, including life-origin theoretical biologists. Certainly any work being done at a subcellular and molecular level would be physical work as physics understands it, wouldn't it? How does physics define work?

3.18.1 The physics definition of work

Work in thermodynamics and mechanics centers around the amount of energy transferred from the environment, or from one body to another. The heat energy given off during the transfer of energy that dissipates is not considered part of that work. The amount of work energy transferred is measured in joules. The rate of transfer is viewed as power, and is measured by watts.

Another way to describe mechanical work in physics is "the force applied to move an object through space along a certain vector (force times distance)." How much force does the wind have to exert on a tumbleweed to blow that tumbleweed up a hill? Physics would consider that "accomplishment" to be "work." Notice how we read into this simple physics problem a formal value judgment that has no place in physics. Our minds subconsciously label the movement of the tumbleweed up a hill as an "accomplishment." Energy was required, to be sure. The wind blew. So what? Why is the new position of the tumbleweed any *better* than its prior position? In an inanimate environment, where a tumbleweed sits has no meaning or value. On top of a hill is no better than the bottom of a hill. Our sentient word "accomplishment" reads into the tumbleweed's movement and new position a lot more than what is justified by pure physics. Our anthropocentric value system affects our supposedly naturalistic understanding of physics all the time. Yet, we are utterly blind to the role our values and knowledge play in our study of physicality. Science is a human epistemology. It is a human mental construction of and about ontological being, not ontological *being* itself. It shouldn't be surprising that any human epistemology, science included, is highly subjective, perspectival, and even prejudicial.

3.18.2 Intuitive, useful work

Useful work is intuitive, functional work. The physics definition of work is completely unrelated to the intuitive, every day, semantic connotation of "useful work" that prevails in human minds. Physicodynamics and physicochemical interactions/reactions have no perception of formal, abstract concepts such as "usefulness." Physicality, apart from agent sentience, is utterly blind and indifferent to notions of "utility" and "value." Physics and chemistry play no direct role in generating "functionality."

The laws of physics and chemistry *can be used by agents* to achieve utility. But that involves steering and control, not just constraint and law. In addition, the value of "functionality" is entertained only within the minds of personal agents. Inanimate nature values and pursues no such thing.

When we talk about protometabolism in a protocell, we are talking about needed *useful work* having to be accomplished.[331] Even the simplest metabolism doesn't just happen. Conceptually complex biochemical pathways and cycles must be directed and steered toward creative metabolic success. In this formal process, energy and the ability to use it are both required. The tendency here is to glibly say, "Usable energy is required." What is required is sophisticated *formal systems* that are able to harness otherwise wasted solar energy, transduce, store, and call it up when usable energy is needed. Sunlight is not "usable" energy. Sunlight is wasted energy from a burning star. The only thing that makes it "useful" are formally controlled Sustained Functional Systems, such as chloroplasts. The genomes and epigenomes of plant cells prescribe and process chloroplasts. They don't just happen.

The prebiotic environment did not perceive or consider formal concepts such as pragmatic goals. One chemical reaction was just as good as any other. No physical interaction or phase change was preferred over any other, except for thermodynamic tendencies in accord with the 2nd Law. This most basic tendency of physicality does not produce physical states far from equilibrium. Only Maxwell's demon's purposeful opening and closing of the trap door between compartments produces states far from equilibrium. Maxwell's demon is an *agent*.

Inanimate physics does not pursue *useful* work. Chemistry does not spontaneously produce a sustained state far from equilibrium. A tumbleweed that just happens to be blown up a hill does not accomplish intuitive, useful work. Any pragmatist should strongly object to naturalists trying to use the physics definition of work to refer to intuitive "useful work." The physics definition of "work" is utterly blind to utility, and cannot possibly address or explain the "useful work" required by cellular metabolism.

3.19 Machines do intuitive, useful work

Machines use energy to perform useful work. Machines such as power tools usually require well-designed and engineered, functional, moving parts. Machines are empowered by mechanical, thermal, chemical or electrical energy. In the case of molecular machines, such as motor proteins, they are usually powered by chemical energy stored in Adenosine Triphosphate (ATP).

What is important is that all machines 1) are Sustained Functional Systems (SFS)[30] (Section 2.4.4 and 3.11) that perform at least one useful task, 2) require an appropriately prepared and supplied energy source to accomplish that task, and 3) must be designed and engineered to utilize that appropriate energy source.

What about "simple machines?" Simple machines may not have moving parts, such as an inclined plane. But an inclined plane (e.g., a hill) does not become a true simple machine until an agent uses that hill to accomplish useful work. Simple machines change the required magnitude of a force, or the vector (direction) of that force. But agents have to use such simple machines to accomplish a desired task before they can consider a machine.

The components of a lever can be found in an inanimate environment. But no levering takes place for any purpose. Even if it did, the environment would not perceive it as "useful work." A prebiotic environment would have no interest in pursuing or preserving levering. It would have no "interests," period.

A wedge must be used by an agent with the intent to pry things apart for a reason. Plenty of prism-like objects can be found in an inanimate environment. None of them is a simple machine until an agent uses one as a wedge to split apart another object to meet some desire or need (e.g., a piece of tree trunk needed to be split for firewood).

3.19.1 How do machines and computers come into existence?

We have absolutely no problem admitting the obvious when it comes to the origin of man-made machines and computational devices, such as computers. We all know full well that no such machines have ever come into existence apart from design and engineering. Physics and chemistry do not produce machines. Design and engineering alone produce such machines. Even then, the chance of malfunction is tremendous.

How many recent automobile recalls have we heard about on the news? What's an automobile manufacturer recall all about? The answer is that less-than-optimal design and engineering becomes evident only through time and use. Despite large teams of highly paid professional designers and engineers, devices fail or do something they weren't supposed to do.

This point seems too straightforward to even mention without feeling foolish. No need exists to defend the logic that machines need careful design and engineering for their production and functional optimization.

But man-made machines are not the only kind of machines that exist. What about the other kinds of machines? Are they any different? How do *they* come into existence?

3.19.2 The simplest cells depend upon millions of molecular machines and nanocomputers

Earlier in Section 3.12.3 we mentioned the motor protein kinesin. This molecular machine is almost more human-like than machine-like. It literally walks along microtubules with two legs while carrying cargo on its back. The load that it carries reminds us of ants. Ants can carry many times their own weight. So can the motor protein kinesin.[332] Kinesin molecular

Section 3: How did primordial prescription, and its processing, arise in a prebiotic environment?

machines form an army of "socialized ants," all making their individual contribution to the "greater societal good" within the cell population of thousands of protein molecular machines and nanocomputer contributors. They line up and carry huge loads of cargo from the nucleus to many destinations in the periphery of the cell. Other motor proteins (e.g., dynein[333-335]) carry loads along microtubules back towards the nucleus.

Numerous more complex motor proteins exist within the cell, the complex conceptual functions of which are essential for cell survival. Molecular machines often have moving parts. Multiple kinds of machines are organized into hierarchies of functional dependence and achievement. Utility of the network reflects a required all-or-none holism. Unless all of the machines are in place and working together, no benefit to the cell is derived.[336]

What is the "purpose" of each? Do we appreciate that "purpose" is a formal term and concept, not a physicochemical reaction? What do those molecules contribute toward holistic metabolism? These questions are the most fundamental of all questions to molecular biology. But, they are not physicodynamic questions addressable by physics. They are formal questions that can only be answered by "systems biology." They are *engineering* questions much more than physics or chemistry questions. The engineering of molecular machines within any cell cannot be reduced to mere "biophysics."[337] Multiple independent parts all work with precision in cooperative holistic schemes.[336]

We said that "systems biology" must address metabolism. Why? A "system" is always formal. A system is always choice-mediated. Everything about metabolism is system-oriented. Metabolism is about the most organized phenomenon known to any field of science. Every pathway and cycle must be integrated into a holistic cooperative concert. The "useful work" of every molecular machine supports the overall formal metabolic schemes of the cell. There are Sustained Functional Systems (SFSs) at every level. The cell itself is a grand meta-SFS.

You will recall in Sections 2.4.4 and 3.11.3 we elaborated on the criteria that define any bona fide "system." Systems always involve formal organization in pursuit of functionality. Remember that a weather front, for example, is not a true "system," because it involves no purposeful choice-contingency. No spontaneously self-ordered physical state qualifies as a true "system." Chaos theory is irrelevant to Sustained Functional Systems (SFS) and systems biology. A Sustained Functional System (SFS) is an effect. An SFS must be caused and directed toward utility by Choice Determinism (CD). Physical Determinism (PD) cannot cause the effect of an SFS.[30]

Not only is the cell itself an SFS (See Section 3.11.3) but cells contain large numbers of organized SFSs within them that collectively contribute to and make possible the overall SFS of the cell.[26,28]

3.19.3 How did so many required subcellular molecular machines and nanocomputers come into existence?

Subcellular nano-molecular machines are large in number and of multitudinous kinds. We know very well exactly what is required for all other kinds of machines to come into existence. Why exactly would the requirements for formation and optimization be any different for molecular machines?[336,338] Do we have some good reason to balk at admitting that molecular machines, like all other machines, need extremely careful design and engineering to form and work well? What logical justification do we have to suddenly dismiss the causative and optimization requirements of any known non-trivial machine? Is it just because they are molecular and smaller? Smaller just makes their design and engineering all the more difficult to achieve. Ask any miniaturization engineer.

The environment in which these nano-molecular machines are found, and where they have to do their job, doesn't change any of the requirements for how they must be produced. Without agency to plan, organize, manufacture and manage their sophisticated functions, subcellular machines are no more likely to just spontaneously appear than human-made machines.[339]

3.19.4 The spontaneous generation of a simple paper clip

Reductionism has paid many dividends in science. Perhaps a much simpler model of origin than abiogenesis might help.

How does a paper clip come into existence?

Let's suppose the happenstantial, spontaneous generation of an optimized paper clip were one's scientific model for the origin of paper clips. Would that model be scientifically *plausible*? Is there any way to *definitively* answer whether that model would be scientifically plausible? Would the hypothesis be testable? Would we be able to construct a null hypothesis that would be potentially falsifiable? For example, "The spontaneous generation of non-trivial paper clips in nature does not happen." This null hypothesis could be potentially falsified by observing a single such occurrence. Perhaps those eager to falsify the null hypothesis could diligently search for any prediction fulfillments of their belief in the spontaneous generation of non-trivial paper clips.

Not finding any empirical evidence in the recorded history of human observation, true believers in this model could still argue that we would need eons of time to observe such an occurrence. That would be a tenable argument. But, given such a contention, that eons of time would be required to observe such an occurrence, belief in the spontaneous generation of non-trivial paper clips would remain just that—a metaphysical belief. It would have no empirical support. It would become a basically non-testable and functionally non-falsifiable belief (some might say, "superstition"). It would in fact have no *scientific* merit.

And, if we are expected to believe that it would take that long for a single paper clip to spontaneously generate, would we also be expected to believe that large numbers of other sizes and shapes and styles of paper clips also formed in that same period of time?

Paper clips are about as simple a "device" as anyone could image. This author feels foolish even mentioning them in such a context. But, people do forget that even paper clips must be carefully manufactured to achieve a consistently optimized product.

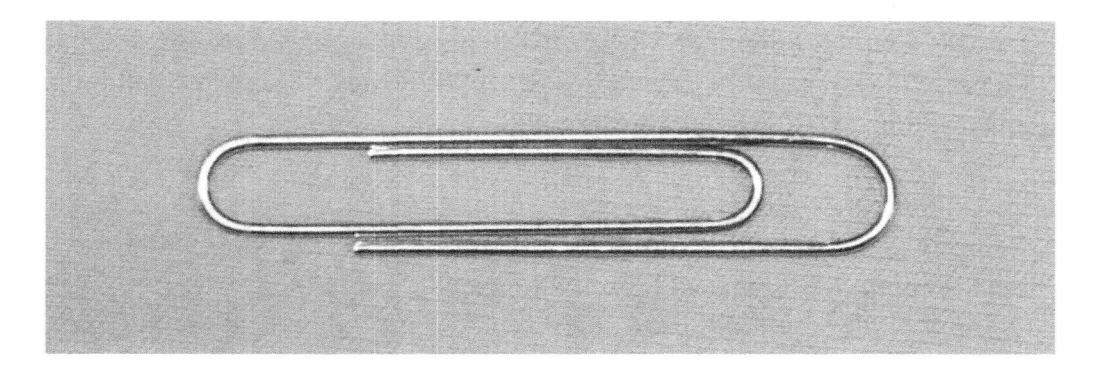

Figure 5. A standard paper clip found in an archaeological dig by an anthropologist.

How quickly we judge a paper clip to be of poor quality. First, "size matters." An optimized paper clip must be adaptable to clasp anywhere from 2 to 20 pages. Some paper clips are more distensible with varying thicknesses of paper. Why are some paper clips so wonderfully compliant without bending in the wrong places? Inferior paper clips bend in the flat dimension with one use, and never function ideally again.

The ideal paper clip is easy to apply, yet not accidentally pulled off the papers. We need to be able to remove that paper clip, however, from the papers almost as easily as we applied it. Such malleability is no accident. The alloy of metal must be precisely specified to optimize its malleability, yet maintain its strength. The alloy must be flexible, pliable, adjustable, compliant, and able to adjust to different thicknesses of paper that will need to be held together. Production of that metal alloy requires continuous high-quality formal *controls*, not just natural constraints.[61] What would be the probability of such an alloy springing spontaneously from iron ore in the ground?

A very long wire of constant diameter would have to be produced by the environment. What is the likelihood of "nature" producing a spool of very long wire of constant diameter?

Then, the wire must be regularly cut at exact intervals. Irregular intervals would result in paper clip wires being cut wrongly, resulting in incomplete paper clips, or paper clips with excess ends. They would all be "mismanufactured" in the production line.

Next, that wire must be bent at all the right places, with no bending occurring at unprescribed points that would weaken the wire or result in a less than optimized shape and pinch for grasping the papers.

Highly optimized paper clips do not leave depressions in the paper. Yet, they tightly grasp all the papers. "That is no accident," as any designer/engineer of manufacturing machinery and paper clips would tell us.

When ideal paper clips are removed from a stack of paper, they don't gouge the surface of the top paper.

How can any single paper clip get optimized—come to meet all of the above criteria?

Could any life-origin analogy possibly be as silly as the origin of a paper clip? Yet, what is the likelihood of such a paper clip spontaneously generating in an inanimate "natural" environment? Can the fixed, forced, redundant, unimaginative laws of physics produce such a paper clip? What is the probability of such a highly optimized paper clip arising by chance? Can mere mass and energy conversions produce even a lousy paper clip? No one has ever observed such an occurrence. What would be the basis for considering such a blind belief in the spontaneous generation of a paper clip "scientific?"

Would a scientist ever suggest that paper clip production *needs* both instructions and carefully controlled manufacturing? Probably. But, they would also argue that the paper clip analogy is not a "physics or chemistry problem." They would be largely right, apart from needing a highly specialized metal alloy. But, why would a chemist feel justified referring us to an engineer rather than a physicist or chemist to investigate the origin and optimization of paper clips? Might there be some "cybernetic" factors in paper clip design and manufacturing that alone can explain paper clip utility and quality?

Why is this fact about paper clips (little more than a "simple machine") any different from the cybernetics that would be required for far more conceptually complex molecular machines?

A naturalistic physicist or chemist cannot remain consistent with his own worldview if he has to "beg off," and appeal to *engineering exceptions* to explain *any* aspect of physical reality, including the manufacture of paper clips. Either physicodynamics is sufficient to explain all pieces of reality's puzzle, or it isn't. If physicodynamics is not sufficient to explain all the pieces of the puzzle, then naturalism as a worldview is bankrupt.

Even if we could explain the origin of instructions for paper clip manufacturing in a prebiotic environment, what would have brought those instructions to fruition? How would the generation of a paper clip have proceeded? What would have processed those instructions? In the end, the existence of the actual physical paper clip is what matters. Does any scientist seriously believe that a paper clip can just pop into existence from mere Physicodynamic Causation (PD) alone?

No one has ever come across a paper clip in the forest floor that was formed by natural process. Archaeologists and anthropologists would universally identify such an object as proof of a previous civilization. We just do not find a metal alloy cylinder of that consistent a diameter and bent at just those points so as to form an ideally functional paper clip. We find iron ore in the forest floor naturally. But we do not find just the right metal alloy with just the right tensile strength and malleability, yet with sufficient rigidity to hold the papers together without ripping or crimping those papers. That kind of paper clip has never been observed to come into existence as a dissipative structure of chaos theory. For one thing, the paper clip is just too sustained a device to be a "dissipative" structure. And it's too exceptionally functional for "natural process" to have prescribed and manufactured it without intent (without Selection *FOR, or in pursuit of*, which evolution cannot do).

We called the paper clip spontaneous generation analogy "silly," if not "ridiculous." Do we have any idea how many orders of magnitude more conceptually complex and sophisticated are the molecular machines found in the most primitive living cells known? Look how many amazing protein molecular machines would be required to assemble in one place at one time for even a ribosome to "self-organize," let alone "metabolize." Most bacteria "manufacture" 3,000 proteins. If a paper clip can't spontaneously generate, why do we so easily choose to believe that thousands of far more conceptually complex molecular machines can spontaneously generate, let alone life? How exactly can we distinguish this faith from superstition?

The First Law of Biology of Pasteur still stands, unfalsified. "All life must come from previously existing life." In other words, "Life cannot spontaneously generate from non-life." It is incumbent upon any scientist who thinks this law is antiquated by evolution to falsify it before ignoring, or trying to dismiss this most fundamental Law of biology. Neither a paper clip nor life has ever been observed to spontaneously generate.

The existence of instructions for how to manufacture a paper clip must never be equated with the existence of a paper clip. Just because a new piece of Ikea furniture comes with assembly instructions does not mean those instructions will ever be used. Most people attempt an intuitive assembly on their own. The results are fairly predictable. Hence, the joke, "When all else fails, read the instructions." Prescriptive Information must be processed. The processing may require more PI than what is being processed.

Thus, not just the instructions for paper clip manufacture are needed. The *means* by which those instructions would have to be implemented is also required. The processing of instructions must itself be prescribed. This processing must be directed to fruition, and the optimized device materialized, to demonstrate prescription of formal function into a physical world.

"Why couldn't an optimized paper clip just appear spontaneously in a prebiotic environment?" is the standard answer of naturalistic science today. "It *could* happen! Molecular evolution could explain it. The probability would be vanishingly small, but the spontaneous appearance of a perfect paper clip in an inanimate environment *could* happen! It would be theoretically possible." The astonishing attitude then emerges from otherwise competent scientists, "Why would anyone dare raise an eyebrow of skepticism over our model? We all know that the spontaneous generation of paper clips is proven scientific fact. Anyone who raises any questions about this model couldn't possibly have any valid scientific credentials. Paper clips could happen! They HAD to have happened, because, obviously, paper clips exist."

"There's iron ore in the ground, isn't there? There are other heavy metals in soil. Earthquakes, landslides, glaciers, a single falling rock, etc. could have bent the forming metal. A paper clip could spontaneously form, given eons of time, couldn't it? We can't eliminate the possibility. The probability might seem infinitely low, even statistically prohibitive. But an extremely low probability never establishes absolute impossibility. It could happen! So why shouldn't we believe in the spontaneous generation of paper clips in inanimate nature?"

The best answer to this idiocy has to do with the scientific measure of *plausibility* of such a model. Plausibility is very different from probability. When the Universal Plausibility Metric (UPM, ξ) of any hypothetical model falls below 1.0, the model is considered to be scientifically falsified.[80,330]

Section 3: How did primordial prescription, and its processing, arise in a prebiotic environment?

Peer review is required to reject any science journal manuscript that proposes such a quantifiably implausible model.

What if there were simpler, more parsimonious models with much higher probabilities of everything occurring harmoniously to produce an optimized product. Shouldn't we abide by Ockham's razor to favor the simplest model that explains the data?

How would other models fare in competition with the above model of spontaneous generation of paper clips? Would other models be "scientific" if they included some mechanism of deliberate, purposeful Prescription and Controlled Processing (CP) of optimized paper clips? "Of course not. That wouldn't be scientific," says the philosophic naturalist. Translated, that means, "It would not correspond to our starting naturalist *metaphysical* beliefs. We have already pre-assumed that no Choice Determinism can be allowed."

As simple as a paper clip is, when all of the independent, non-conditional probabilities of a spontaneous paper clip formation are multiplied out, the probability of a spontaneously optimized paper clip occurring in a prebiotic environment becomes quickly statistically prohibitive. And then, there are a lot of conditional and non-independent probabilities on top of that. With a lot of work, crudely measurable probabilities could probably be calculated for each necessary and sufficient condition required for spontaneous paper clip formation in an inanimate environment.

Let's suppose paper clips somehow could reproduce themselves. There might be some hope of passing along the capability of reproduction methodology. In that case, the wheel wouldn't have to be re-invented with each new paper clip formation. Of course, there would have to be some means of "writing" Prescriptive Information (PI) (instructions for making paper clips) independent of the actual physical paper clip itself. A mere pile of paper clips wouldn't guarantee propagation of paper clips any better than one paper clip. And those instructions would have to be recorded in a physical medium of retention and transmission. There would have to be some representational Material Symbol System (MSS) for coding those instructions. A set of arbitrarily-chosen semiotic or cybernetic *rules* (not laws) would have to be followed. Otherwise, interpretation and processing of the token (physical symbol vehicle) syntax would be impossible. But, even if the instructions could somehow be recorded and read by an inanimate environment, there would have to be sophisticated mechanisms and devices to process that PI. Next, there would have to be a crucible, the right temperature and mixing machine of some sort to blend the right combination of melted metals. Just the right temperature would be needed for just the right amount of time. Then molding would be needed to pour the metal into, or some sort of stretching device that could string the forming metal strand out into a wire. How would constant diameter be achieved in a prebiotic environment? We have no idea how any of these "problems" "could be solved." But, if we were rationally consistent naturalists, we would have no choice but to argue, "It could happen!" We might also be illogical enough to argue, "We *know* it happened, because paper clips exist!"

Suppose an anthropologist or archaeologist finds a standard paper clip buried four feet down on a dig (**see Figure 5 Section 3.19.4**).

What would that anthropologist or archaeologist conclude scientifically about the find? Is the paper clip evidence of a prior civilization? Or would that scientist conclude it was a natural, spontaneous product of inanimate nature? How would science go about determining which hypothesis more closely corresponded to objective reality?

Any respectable anthropologist/archaeologist knows immediately when obvious products of agent contrivance are being examined. The spontaneous generation of a paper clip is just too implausible to entertain as a responsible scientific model of "natural" occurrence from physical and chemical laws.

Should we respect the intellect of someone who persistently argues that paper clips just happen in nature, given enough time? Paper clips just molecularly evolve into existence?

Time is not a cause of any physical effect, let alone formal abstract concept, engineering and function. A lot of time does not generate paper clip instructions, processing of those instructions, optimized metal alloys, wire-making machines, benders and cutters, any better than a short amount of time.

What is far sillier than the paper clip analogy is a faith system that subscribes to belief in the spontaneous generation of life in an inanimate environment. This is true in any amount of time within naturalistic cosmological age estimates. 14 billion years converts to a surprisingly small number of seconds when compared to the probability of a simple optimized paper clip "self-organizing" from iron ore in the ground. Even more ridiculous than that is the contention that such a faith system is scientific.

3.20 Probability vs. Plausibility

It is certainly reasonable to contend that even extremely low probabilities are still "possible." "It *could* happen!" But, in science, that is not the question. The question is:

> "What is the *plausibility* of a model with such an extremely low probability?" This question has direct bearing on the "usefulness" of any scientific theory.

Science has a definitive way of quantitatively evaluating the *plausibility* of low probability scientific hypotheses, models and theories. It is called the Universal Plausibility Principle.[80,84,330] The Universal Plausibility Principle permits definitive falsification of hypotheses with a Universal Plausibility Metric (Xi, or ξ) that falls below 1.0^{80} A scientific theory cannot be eliminated simply because of an extremely low probability of occurrence. It *can* be eliminated from scientific consideration, however, if its Universal Plausibility Metric ξ is < 1.0.

It is not always possible to calculate reliable probabilities for certain proposed scenarios, especially when a single historical event (or very few events) are championed. Often a crude low probability can be reasonably estimated for that scenario, however. When multiple independent, non-conditional events are all required to get a certain final product, probabilities must be multiplied. The probability of the spontaneous generation of a simple paper clip that meets all of the above requirements, when all crude probabilities are multiplied out, can easily become so daunting as to result in a UPM ξ far, far below 1.0. Given the probability bounds of cosmic age, the number of elementary particles in the cosmos, and even the most far-out imagination,[286] the UPP is grossly violated. The notion of spontaneous generation of optimized paper clips can be unquestionably and quantitatively falsified. If this is true of paper clips, it most certainly is true for the spontaneous generation of life.

3.21 Bio-informational Turing tapes and machines

All analogies break down eventually. But much benefit can usually be gained through analogy applied to specific points. DNA, and especially edited mRNA viewed as Turing tape, certainly has at least some analogous benefit. And, of course, that tape must be processed by some semblance of a Turing "machine."

What exactly is a theoretical Turing machine and tape?

Something is a Turing machine if it has a 'tape' that extends infinitely in both directions, with the tape subdivided into identical adjacent squares, each of which can have written on it one of a finite alphabet of symbols (usually just zero and one). In addition, a Turing machine has a 'tape head,' that can move to the left or right on the tape and erase and rewrite the symbol that's on a current square. Finally, what guides the tape head is a finite set of 'states' that, given one state, looks at the current symbol, keeps or changes it, moves the tape head right or left, and then, on the basis of the symbol that was there, makes active another state. In modern terms, the states constitute the program and the symbols on the tape constitute data.[340]

Essentially, all a universal Turing machine does is: read instruction, execute it, read instruction, execute it, read instruction, execute it, etc.[340]

We must acknowledge a sort of division of labor that exists within the operation of a universal Turing machine and an analogous ribosome. Processing mechanisms have to act on the Turing tape. These processing mechanisms cannot be reduced to the programming or data on the tape alone. More PI is needed than just the programming and other data on the Turing tape itself.[23,24] PI is required to generate a Turing machine (e.g., ribosome), to instruct, and to run it.

Theoretical universal Turing machines model algorithmic computation itself, not just computers. Much of molecular bio-function within the cell consists of true algorithmic computation. Chance and necessity/mass and energy, are incapable of algorithmic computation. The latter is fundamentally formal and choice-contingent.

The Turing tape of an edited mRNA not only contributes to the program, it also provides prescriptive *data* as to which particular polyamino acid sequence and protein the ribosome Turing machine will produce. Not only does each "state" have to be read and executed by the reading head of the ribosome Turing machine, but the sequencing of codons determines which protein will be "computed," or manufactured. In addition, which one of the multiple redundant codons is used to prescribe a certain amino acid determines protein folding and function.[253,254]

The reason the Turing machine and tape analogy is so relevant has to do with 1) the Church-Turing thesis, and 2) the meaning of "computation." Any function is considered algorithmically computable if and only if it is computable by a Turing machine. An edited mRNA and the reading and execution of that "tape" performed by ribosomes are legitimate examples of "algorithmic computation." Not only that, but they provide example of the most highly optimized algorithmic computation known to science and engineering. New overlapping layers of PI are being discovered faster than what the biological community can keep up with.

The polycodon sequence on the mRNA tape, for example, not only prescribes the sequencing of amino acids. It also prescribes translational pausing (TP) during protein manufacture within the ribosome. TP in turn determines the protein's three-dimensional structure as it emerges from the back end of the ribosome.[253,254] The instructions for TP and protein folding are contained in the supposed "degeneracy" of the codon table. The redundancy of more than one codon prescribing the same amino acid, in other words, is anything but "degenerative." It is no accident that *a certain one* of the redundant codons is needed to prescribe a certain amino acid. More is involved in the programming of a protein and its folded structure than just the prescription of its amino acid sequence. The correct codon must be used from among several codons, any one of which could prescribe the correct amino acid. This is an example of the multi-layered, or multi-dimensional, Prescriptive Information (PI) found in the edited mRNA strand. There is no way that chance and/or necessity, mere mass and energy, could have prescribed such multi-layered, multi-dimensional PI instructions into the same mRNA strand.

When Bill gates made the following statement, he had no idea just how multi-layered biological PI was. The above TP knowledge, for example, did not even exist at that time:

> *"DNA is like a computer program, but far, far more advanced than any software ever created."*[341]

DNA is far more sophisticated than any artificial Turing tape. And the ribosome is far more sophisticated than any artificial Turing machine. DNA is a physical biopolymer. But the physicality of this molecule is not what Bill Gates was referring to when he called it *"far more advanced than any software ever created."* Software is instantiated into a physical medium of retention. But the software *itself* is not physical. What exactly is software?

Software is a string of computationally successful, wise *programming choices*. We pay good money for reputable software; we want our money back when it functions poorly because of "bugs." Bugs are less-than-efficacious programming choices at the true decision nodes that comprise the program.

The choices that generate "advanced software" are wise nonmaterial selections from among multiple options. Notice that although the options from which to choose may be physical (tokens; physical symbol vehicles), the choices themselves that select those physical objects are nonphysical. It is nothing short of mind-boggling how so many otherwise brilliant investigators have so much trouble making this simple distinction. The persistent blurring of these two crystal-clear categories precludes progress into understanding the nature of biological Prescriptive Information (PI).[20,71] The physical medium merely provides a registry of choice results. If the contention is that mind and intent are ultimately nothing more than epiphenomena of physicality, it is incumbent upon the materialist to explain how chance and/or necessity, mass and/or energy, can generate Choice-Contingent Causation and Control. So far, no naturalistic neurophysiologic model has come close to falsifying either Cartesianism, or the less philosophic and more scientific F > P Principle (Formalism > Physicality Principle)[19] (See Section 3.16).

3.21.1 Software prescribes not-yet-existent formal function into existence

"Cybernetic" means "steered, directed, programmed, controlled, regulated, even designed and engineered, toward future functional/pragmatic success." Controls precede and produce utility. The fittest formal function cannot be selected by an environment before that function exists. It must first be prescribed into existence, and then managed or controlled to maintain its efficacy.

Cybernetics is always *choice-contingent*. Cybernetics arises out of *Decision Theory*, not stochastic theory. This is the reason Shannon so-called "information theory" fails so miserably to define or measure real information. Shannon theory is based solely on stochastic analysis. It acknowledges only constraints and statistical boundaries, not formally steered controls. Mere probabilistic combinatorial theory cannot begin to explain cybernetics or its efficaciousness. Neither can Monod's "Necessity."[40]

Fixed, forced law cannot participate in Decision Theory any better than chance-contingency. Decision theory requires freedom (contingency) from law-like determinism. But, mere contingency is not sufficient. Decision theory also requires choice-contingency, not chance-contingency.

The essence of Prescriptive Information (PI) is the efficacious choice made at each decision node. DNA is an instructive string of programmed decision nodes. Genomic and epigenomic configurable switch settings integrate circuits. Biopolymer messenger molecules instruct, prescribe and process life into existence. PI operates the opening and closing of its configurable switches.

Not all Functional Information (FI) is prescriptive (PI). Some FI is merely Descriptive Information (DI). Subcellular life, however, depends up PI, not DI.

Intuitive, semantic information is always *about* something.[342-346] There is nothing physical in "aboutness." "Aboutness" is altogether formal, not material.

Intuitive, semantic information provides a relative certainty, rather than mere Shannon uncertainty, about what is going to "work" (or be useful). Cybernetics establishes Choice-Contingent Causation and Control (CCCC) to be an empirical and rationally definable third legitimate fundamental category of reality in addition to chance and necessity. Cybernetics proves Choice Determinism (CD) and its all-pervasive computational functional fruit to be an undeniable aspect of the objective real world in which we find ourselves. "Choice matters" because choice is a true cause of real effects. Mathematics, logic theory, and science cannot be practiced without "choice with intent." No one, regardless of metaphysical commitments, can successfully argue that mathematics, logic theory and the scientific method don't matter, or that the choice-contingency and choice determinism (CD), upon which they all depend, isn't real.

3.21.2 Venter's "programming of a digital organism"

Let us return for a minute to the Bill Gates quote about DNA being the most sophisticated software known. We might quip, "Bill Gates is not a biologist. What does he know about DNA and molecular biology?" Perhaps referring to DNA in the context of containing instantiated "advanced software" is just an overdrawn, bad analogy from a non-biologist.

Craig Venter, however, led the way in mapping the human and other genomes using computer science rather than the more traditional means employing physical DNA. Why and how would computer science have been so helpful in identifying and analyzing DNA instructions? His emphasis was on formal, nonphysical algorithms and computation outside of the cells themselves. His digital and representational approach to the Human Genome Project outpaced in many cases Francis Collins' work with physical DNA mapping of reference sequences. Venter's more cybernetic approach worked for one reason: *it corresponded to the way things actually are within the cell.* Genomic and epigenomic prescription are not just *like* a computer. The cell actually consists of thousands of nanocomputers, and other molecular machines. All these devices and processes cooperate to achieve a common purpose.[26,347] Millions of parallel, integrated computations are going on every nanosecond in any cell. All these computational tasks not only work simultaneously toward common goals; they all contribute to the one meta-goal of being and staying alive. The cell is not just analogous to cybernetics. The cell *is* cybernetic. It is controlled by edited, very real linear digital software, by computational processing devices, and by organizing and governing formal operating systems that are highly optimized. The alternative splicing and switching on and off of genes are highly cybernetic.

We cannot dismiss Venter's knowledge of biology as we do with Bill Gates. Venter's team is sometimes credited with being the first to "program a synthetic organism."[348] The work involved techniques similar to those used in cloning. Venter's team used digital and cybernetic engineering to insert copied genetic instructions from one species into an existing cell of a different species. The artificial instructions were able to take over the cell. Venter's digital programming provided *formal instructions and controls*, not mere physical constraints. His work proved that life really is cybernetic in nature.

It is not surprising therefore, that Venter's perspective on what makes life possible (and unique) centers on information technology. He sees the cell and its function in terms of hardware and software engineering.

It certainly changed my views of definitions of life and how life works . . . Life is basically the result of an information process, a software process. Our genetic code is our software, and our cells are dynamically, constantly reading that genetic code, making new proteins, and the proteins make the other cellular components.[315,349]

Continuing, Venter said:

This is now the first time where we've started with information in the computer, built that software molecule, now over a million letters of genetic code, put that into a recipient cell, and have this process start where that information converted that cell into a new species.

. . . A powerful tool for designing what we want biology to do.

It's pretty stunning when you just replace the DNA software in the cell, and the cell instantly starts reading that new software, starts making a whole different set of proteins, and within a short while, all the characteristics of the first species disappear, and a new species emerges from this software that controls that cell going forward.[315,349]

Speaking from the same podium from which Schrödinger delivered his famous "What is Life?" lecture 69 years earlier, Venter essentially argued that life's symbol systems and coded instructions are not just linguistic and computer-like, but are in fact literal cybernetic systems:

> All living cells that we know of on this planet are 'DNA software'-driven biological machines comprised of hundreds of thousands of protein robots, coded for by the DNA, that carry out precise functions. . . . We are now using computer software to design new DNA software.[350]

Clair O'Connell, writing for *New Scientist*, covered Venter's lecture with phrases like "artificially synthesizing DNA to reboot cells." Venter argued that the digital and biological worlds are becoming interchangeable. Scientists no longer exchange biological material. They simply send each other the *digital information* needed to make that biological material in the lab. Says O'Connell:

> Venter also outlined a vision of small converter devices that can be attached to computers to make the structures from the digital information - perhaps the future could see us distributing information to make vaccines, foods and fuels around the world, or even to other planets.

> But perhaps the most intriguing anecdote Venter shared was his description of how his team 'watermarked' their synthesized DNA with coded quotations from James Joyce, Robert Oppenheimer and Richard Feynman, only to learn that they had included a mistake in

the Feynman quote. Venter's rather airy description of how they just went back in and fixed it drove home just how far we've come in being able to understand, and even manipulate, our own DNA molecules."[350]

Here perhaps, O'Connell seemed to have missed Venter's main point a bit, overemphasizing, and perhaps confusing our DNA physical molecules, with the nonphysical programing that is instantiated into life via the sequencing (the syntax) of nucleotide tokens. Our DNA nucleotides are used in a Material Symbol System (MSS)[43,44,59] to prescribe amino acid sequencing in proteins, along with transitional pausing that controls protein folding.

Even before addressing the problem of ontological Prescription, the chemical roadblocks to making DNA and RNA in modern labs are well known. Those biochemical problems are far, far worse in any prebiotic environment. This has forced naturalistic science into hypotheses involving exotic, creative new analogs of RNA that might have been easier to form, and more stable in prebiotic nature.[264-267,351,352] Very recently, a semi-synthetic organism with an expanded genetic alphabet was "created" through highly focused "engineering."[353] "Engineering" is not a function of chance and necessity, mass and energy, but is entirely choice-contingent. The resulting bacterium is the first organism to propagate stably an expanded genetic alphabet. So far, RNA polymerases are not able to transcribe the novel DNA into mRNAs. But there is no reason to doubt that Choice Determinism (CD) could eventually engineer that.

Even before Venter's lecture, Richard Dawkins pointed out:

"If you want to understand life, don't think about vibrant, throbbing gels and oozes, think about information technology."[354, pg 112]

Donald E. Johnson, with PhD's in chemistry, information science and computer science, also sees life in terms of cybernetic programming and

Section 3: How did primordial prescription, and its processing, arise in a prebiotic environment?

information processing.[27,28,315,355] Analyzing Venter' work and comments, Johnson says:

> The new genome was engineered using computers 'starting from digitized genome sequence information.' This verifies the digital computing nature of life. The complexity and specificity of life's information is highlighted by a quote from Venter:
>
>> 'Obtaining an error-free genome that could be transplanted into a recipient cell to create a new cell controlled only by the synthetic genome was complicated and required many quality control steps. Our success was thwarted for many weeks by a single base pair deletion in the essential gene dnaA. One wrong base out of over one million in an essential gene rendered the genome inactive.'[355]

Johnson continues,

> One wonders how 'nature' could have achieved such specificity with no intelligent direction. One of the things this research supports is the idea that (at least for the two bacteria involved) life uses common operating systems, programming languages, and devices (otherwise the programs for one machine wouldn't execute on another).[355]

Stanford-educated chemist Nivaldo Tro points out that near the end of his life, Stanley Miller remarked:[356, pg. 135]

> "The problem of the origin of life has turned out to be much more difficult than I, and most other people, envisioned."

Professor Miller stated basically the same thing to me personally in several one-on-one conversations near the end of his life.

Leslie Orgel was even more forceful, even within peer-reviewed literature:

> "In my opinion, there is no basis in known chemistry for the belief that long sequences of reactions can organize spontaneously—and every reason to believe that they cannot."[357]

Like Stanley Miller, near the end of Leslie Orgel's life, he became more pessimistic about either an RNA analog world scenario, or a Metabolism-First model being the least bit plausible.[358] Many of Polanyi's opinions wound up being affirmed by the most optimistic and prominent of naturalistic life-origin researchers:

> As the arrangement of a printed page is extraneous to the chemistry of the printed page, so is the base sequence in a DNA molecule extraneous to the chemical forces at work in the DNA molecule. It is this physical indeterminacy of the sequence that produces the improbability of any particular sequence and thereby enables it to have a meaning—a meaning that has a mathematically determinate information content.[359]

Here, Polanyi was a bit confused. This is quite understandable for his time. "Meaning" is not mathematically determinate. The rules for equation manipulation, for example, are ultimately linguistic. And intuitive, semantic information cannot be measured with any derivative of stochastic theory (e.g., Shannon Uncertainty, mutual entropy included). (See Section 2.16)

Paul Davies views the cell as "an information-processing and replicating system of astonishing complexity."[360] Davies continued:

> DNA is not a special life-giving molecule, but a genetic databank that transmits its information using a mathematical code. Most of the workings of the cell are best described, not in terms of material stuff—hardware—but as information, or software. Trying to make life by mixing chemicals in a test tube is like soldering switches and wires in an attempt to produce Windows 98. It won't work because it addresses the problem at the wrong conceptual level.[360]

Davies, however, has no idea how physicochemical interactions could have produced life's instruction set and algorithmic processing of that programming.[361]

3.21.3 Prescription, with its programming and processing, are the most defining characteristics of life

Many would refer to Prescription and its Processing (P & P) as "Information technology." DNA is probably the most dense and stable information medium known to science. DNA has a data density of one million gigabits per cubic millimeter. In a SCIENCE paper entitled "Next-Generation Digital Information Storage in DNA,"[308] Church and his associates encoded a 5.27 MB book using DNA microchips. This amounts to a 53,000-word book, including 11 JPG images and one JavaScript program, all encoded into DNA. Goldman, et al.[362] echo the same findings.

Philosophic materialists go through unbelievable contortions of rationalization and obfuscation trying to deny the digital and cybernetic nature of life.[363] Programming cannot be reduced to physicochemical interactions.[364] Genomics and epigenomics demonstrate the most sophisticated examples of digital prescription, regulation and control known to computer science. The metabolic processes of life continue to inspire creative computing.[365]

Why do Dawkins, Venter, Johnson and Church find it necessary to cite "*information technology*" as the single most distinguishing characteristic of cellular life? We normally think of "technology" solely in terms of human inventiveness. But life preceded humans. Naturalistic science argues that inanimate nature spontaneously produced life, and that life's cellular information technology ultimately produced the humans that, in turn, produce human technology. The scientific problem of life origin, therefore, centers on the question of *how* inanimate molecules in nature could have given rise to the *information technology* needed for life to come into existence.[364] Thus, the origin of life problem focuses on the problem of the origin of literal information technology as observed in every living cell. What do we do with the clearly observable fact of subcellular *programming,* and the fact that *both programming and the processing of that programming require wise purposeful choices*?

When a programmer sits down to program a computational program, no final computational function exists yet. The programmer must make large numbers of wise purposeful programming choices at true decision nodes that will only later, in the future, result in computational success. The programmer must Select *FOR (in pursuit of)* potential function rather than Select *FROM AMONG already existing fittest function.*

It is no different in molecular biology. Evolution is never anything more than Selection *FROM AMONG*. Evolution is never Selection *FOR (in pursuit of function)*. Yet Selection *FOR (in pursuit of)* is the key, not only to abiogenesis, but to every single step-up in sophisticated genomic and epigenomic instruction of every single higher life form. The genome and epigenome must be programmed BEFORE the next more sophisticated organism can come into existence.

What is the mechanism for "PRESCRIPTION" of sophisticated function? All empirical and rational evidence in the history of humanity testifies to the fact that prescription of sophisticated function has never arisen independent of Choice Determinism (CD).[41,42,61]

3.22 Life is a programmed, cybernetic, highly-regulated, computational *process.*

In a Nature paper by Malyshev, et al, they state:

"Organisms are defined by the information encoded in their genomes, and since the origin of life this information has been encoded using a two-base-pair genetic alphabet (A–T and G–C)."[353]

Section 3: How did primordial prescription, and its processing, arise in a prebiotic environment?

The naturalistic worldview dictates that the genomic and epigenomic programming of life just spontaneously generated from mass and energy, chance and necessity. Realizing the absurdity of arguing that "chance did it," naturalism looks desperately to some yet-to-be discovered new law that will one day explain the programming of life. But, no fixed, deterministic law will ever be able to explain programming or prescription of any kind. This book will show that without programming, life could not possibly have arisen. This determination will not be probabilistic. It will result from a definitive scientific falsification of the absurd notion that even a primordial cell could have self-organized.[80,314,330]

A more appropriate subject of interest in life origin science than "information" is the phenomenon of "programming."[26-28,347] We have learned a great deal about genetics, genomics and epigenomics. The ENCODE initiative in biological science has only affirmed all the more that life is programmed with configurable switch-settings at every turn. Life could only have arisen out of *Choice Determinism (CD)* at true decision nodes, rather than redundant, boring, unimaginative Physicodynamic Determinism (PD). Such determinism eliminates contingency, and therefore bifurcation points. Bifurcation points are not synonymous with decision nodes. Technically, for even bifurcation points to exist, freedom from PD must exist (contingency). But, from a naturalistic perspective, freedom from law-like determinism leaves only one remaining category: chance-contingency. Chance has never been observed to program any non-trivial computational success. In fact, chance is not a physical cause of anything. Chance is nothing more than a *descriptive* probabilistic concept. The bottom line is this: Neither chance nor necessity can program computational success. Neither chance nor necessity "know" what "computation" or "success" are. Both concepts are formal, not physical (See Section 2.7 and 2.8).

Intuitive, useful work (not the "work" of physics) is formal. So are function, integration, and organization of biochemical pathways and metabolic schemes. They are not mere physical reactions or phase changes. Steering toward metabolic success is abundantly evident even in the simplest life forms. Nothing exists in inanimate prebiotic nature to steer physicodynamic interactions towards formal functional success. The programming, computation and other aspects of "information technology" found in living things are unique in nature.[284]

3.23 What could possibly produce subcellular "information Technology"?

Cybernetic means controlled, not just constrained. Control and regulation are forms of prescription. Prescription is mediated through how-to instructions, programming, and engineered processing of that programming. Instructions must be processed. Processing itself requires Prescription.[23,24] Even primordial life would have required prescriptive programming and the processing of that Prescription.[26-28]

> No single word relative to biological investigation in the last five years has dominated the scene more than the word, 'regulation.' All known life is cybernetic. Cybernetic means steered, controlled and/or regulated with purposeful intent. What naturalistic life-origin science has never been able to explain is how inanimate chemistry could have formally integrated components and circuits into such holistic organization. Reactions are guided through pathways and cycles into highly conceptual, abstract, functional metasystems. Sophisticated components are manufactured and assembled into molecular machines and nano-computers[26,27] that all cooperate to achieve goals normally considered to be formally transcendent to mere physical interactions. Life is a programmed and pragmatic enterprise. Mere physicodynamic constraints have no motives. They have no formal agendas. They do not pursue functional success.[29]

The only logical escape from having to deal with the necessity of Prescription and its Processing (P & P) would be to:

1) Redefine life to something different from empirical life, or
2) Argue that initial life was different from current empirical life.

Section 3: How did primordial prescription, and its processing, arise in a prebiotic environment?

In an effort to avoid the nasty problems of needed P & P, both avenues (redefining life, or arguing that initial life was different) have been pursued exhaustively in life-origin literature for many decades.

The first approach involves trying to define down life to the point where it needs no controls. Metabolism-first models of life-origin seek to bypass any need for top-down Prescription and its Processing (P & P).[336,338,366] They argue that initial life was unprescribed, happenstantial, self-organized, self-assembled, spontaneously metabolic, and relentless in its ascent toward ever-increasing conceptual complexity.[367] Chance and Necessity are believed to be the sole causative factors. This faith system is utterly bankrupt scientifically. It is pure superstition.

Probabilistic combinatorialism, plus the laws of physics and chemistry, together lack any steering of physicodynamic interaction towards formal organization or sophisticated function.[282-287,368-377]

The intent of programming is to organize and algorithmically optimize computational function. A process must be laid out leading to an executable algorithm.[378] Successful prescription of computational function must be realized (empirically observed) before we can call a string of instructions a "program." This necessitates including successful processing within the definitions of "prescription" and "programming."

Another major problem is the Second Law's relentless tendency towards disorganization. (See section 3.11.5)

The vast majority of an organism's configurable switches have already been set prior to natural selection taking place. Their circuits are already integrated, their computations are already halting, their structure is already organized, and they are already-living before environmental selection can take place. Natural selection is nothing more than the differential survival and reproduction of the best already-programmed, already-living phenotypic organisms. Evolution is Selection *FROM AMONG* populations of living organisms. It is not *Selection FOR (in pursuit of)*. The environment and evolution have no aims or goals.

When an epigenetic switch is flipped (e.g., methylation of a certain cytosine nucleotide), functionality can immediately change. Thus, we might think that we can equate the selectable phenotypic change (e.g. superior fitness) with Selection *FROM AMONG* methylation vs. no methylation. How easily we forget how many different component parts, biochemical pathways, integrated circuits, formal linguistic-like rules, specific nucleotide and amino acid choices, messaging and transport systems, and structural organizations have to be in place and fully organized for a simple methylation to have any switching effect on that circuit, let alone selective benefit. The effect of cytosine methylation has to be programmed to generate formal functional controls. Accidental methylations would never support local, let alone holistic, metabolism.

The first protocell and primordial cells could not possibly have come into existence without formal controls at the subcellular level. Any happenstantial progress would have been extremely limited in any Metabolism First model. That progress would have needed to be retained. Re-inventing the wheel with each new pseudo cell formation would not have worked. Even a minimal cell would have been statistically prohibitive as a first-time happenstantial occurrence. Some heritable system would have been required very early on. That means some Material Symbol System (MSS) that could prescribe phenotypic success had to exist. This MSS could then mutate a duplication somewhat independently of that phenotype, especially in neutral zones. This separation of genotype from phenotype would have been needed almost from the very beginning for evolution theory to have any hope of explaining what it purports to explain. Even then, no explanation exists for how chance and necessity could have programmed computational success and three-dimensional binding function. The most extreme reductionism of primitive cellular physiology fails to render Metabolism-First models plausible. The needed organization is just too extensive, conceptual, and sophisticated for hypercycles or chemotons to integrate into holistic metabolic schemes.

The physical symbol vehicles (tokens) would have had to be selected in advance of any computational or physical structural metabolic utility. Ribonucleotide sequencing in some RNA analog, for example, would have been rigidly bound chemically prior to the realization of any integrative bio-function. Ribonucleotides would have had to be Selected *FOR (in pursuit of)* potential function, in advance of polymerization and bio-function, not just after the fact of function (Selection *FROM AMONG*.

Logical consistency is a requirement of logic theory. But logic theory (like mathematics, language, and cybernetics) is formal, not physical. If the worldview of materialism/physicalism/naturalism accurately corresponds to the way reality actually is, formalisms would not exist. Physicalism maintains that everything that exists is physical. To defend the idea of physicalism, one must resort to nonphysical logic theory. The more one argues in defense of physicalism, the deeper a hole one digs for oneself. One has to employ nonphysical formalisms to support the contention that physicality is all that exists. This is a suicide mission. Even if one succeeded, one would only return to physics—the study of physicality—which is 70% formal mathematics. Anything in physics that is not mathematical requires language to convey, which is also formal rather than physical. The scientific method itself is formal rather than physical.

Materialism/physicalism is dead. No "ism" is physical, including naturalism.

3.24 Duplication plus Variation

"Duplication plus variation" is the backbone of evolution theory. DNA sections are duplicated. Then one of the copies mutates to form new and supposedly better instructive segments. This, along with recombinations, is believed in macro-evolutionary theory to be the major source of superior new genetic and genomic programming for every single new genus and family of organisms.

Has any macro-evolutionist ever stopped to ask the simple question, "Duplication of WHAT?" "Duplication plus variation" always just presupposes the prior existence of high quality Prescriptive Information (PI)^2 that can be duplicated. Where did this high quality PI come from? And, why is it *worth* duplicating?

"Duplication of existing PI" never addresses, let alone answers, what "wrote" the initial PI. What made the initial programming decisions that produced any organism in the first place, let alone "the fittest" organism. No empirical evidence or deductive logic exists in support of the notion that physical law or random interactions could generate formal instructions of any kind. No cause from within pragmatically blind physicodynamics is citable that can explain the generation of the formal symbol systems used to record and communicate bio messages by even the simplest life forms. In addition, the specific programming itself, using that symbol system, is never explained.

It is widely appreciated by information theorists that no new information is generated by "duplication." This is true even with "bits" of Shannon Uncertainty. The only value of duplication in "duplication plus variation" as a hypothetical source of new genetic instructions is to provide additional copies of already existing Prescriptive Information (PI). Existing PI could then vary without disturbing the needed pre-existing Prescription that was programmed by the initial strand.

Also presupposed is that "variation" is going to be beneficial. SEZ WHO? "Change" and "alteration" are not synonymous with "better." What exactly is the basis in empirical science for mere variation being a source of superior new programming? Usually, about the best we can come up with as a beneficial mutation is the sickle cell anemia mutation. This mutation causes defective erythrocytes (red cells). The defect secondarily leaves the red cells more resistant to the malaria parasite. No wonder! The erythrocyte is so screwed up (crenated) as a result of "variation" (typographical error; mutation) of what *were good instructions*, that not even the *Plasmodium* malaria parasite wants anything to do with the now defective red cell! Ask any sickle cell anemia sufferer if the mutation that produced his misery is "beneficial." Most would rather have to deal with malaria than sickle cell anemia!

It is widely acknowledged that probably only one mutation in a million is beneficial. The trouble is, we can't even seem to document very well that one per million. We can't prove that it really was a definite improvement in programming and computational prowess.

The variation that is added to duplication is nothing more than noise pollution of any existing Prescriptive Information in the strand. So, we have the supposedly scientific hypothesis that duplication (no new information) plus variation (noise pollution) produces sophisticated new programming instructions! How could the otherwise most brilliant minds in biology possibly buy into such utter nonsense?

We need to entertain, as a null hypothesis, that duplication plus variation *cannot* write new non-trivial Prescriptive Information. Could such a null hypothesis be falsified? Yes, it could be. All we would have to do is to produce a single instance of spontaneous "duplication plus variation" producing new non-trivial PI. We might be able to find such evidence within the *resorting* of existing instructive segments. Even then, however, improvement in the integration and overall computational effects of such *resorting*, such as "crossing over," is not as easy to prove as we might think. Inconsequential (non-deleterious) "neutral" mutations are common, but computationally superior mutations are not.

Mobile genetic elements are a different story. In these instances, already-existing PI can be introduced into a genome from other programmed sources. The existing affected genome is not reprogramming itself in any way.

It is finally becoming more acceptable within scientific literature to admit that the answer to the question, "Duplication of what?" is Prescriptive Information (PI). Why do we find it necessary to capitalize this term? The answer is that the scientific community's understanding of "information" is embarrassing. The definition, connotations and usage of "information' in peer-reviewed literature generates untold mental mush. Information is thoroughly confused and conflated with Shannon Uncertainty and Reduced Uncertainty (poorly termed "mutual entropy"). Investigators almost universally fail to realize that even Reduced Uncertainty (R) is still a measure of nothing more than Uncertainty. In addition, the injection of Bona fide Functional Information (FI) was required to reduce that Shannon Uncertainty. Without supplying this external FI, uncertainty could never have been reduced.

In the absence of Choice-Contingent Causation and Control (CCCC), variation is nothing more than noise. Noise can't program anything, let alone computational success, computer architecture, system rules, or highly integrative and interdependent metabolic schemes. No reason whatever exists to think that mere "variation" of existing PI could generate sophisticated new programming schemes. No empirical verification or prediction fulfillments exist in support of such a notion. Most variation, especially in duplicated PI, is neutral (silent and unselectable). In other words, the mutation occurs in a non-critical region of the DNA molecule. It neither hurts nor helps DNA's coded instructions. But neutral mutations are usually the best "variation" can achieve. Kimura[244] couldn't even graph beneficial mutations on the positive side of his graphs of neutral mutations. Virtually all mutations that are not neutral are deleterious because valuable instructions are scrambled by the equivalent of typographical errors.

In addition, what used to be viewed as neutral mutations are fast being discovered to be anything but neutral. They may be neutral in so far as the prescription of a certain protein's amino acid sequence. But, they are not neutral with respect to non-protein-coding prescriptions and epigenetic switching controls.

Technologies arise only out of Decision Theory, not Probability Theory. Probability theory is descriptive, not causative. Decision theory is determinative. Decision Theory possesses the capability of *prescribing* new organization and formal function. Mere variation does not.

Technologies cannot arise out of fixed natural law, either. Purposeful choices have to be made out of freedom from physical law, at real decision nodes, in order to organize objects and steer events towards sophisticated functional success. Chance and necessity are the only two categories of reality affecting non-teleonomic "variation." Neither coin flips nor physical law can program cybernetic or semiotic systems. Life is undeniably dependent upon both cybernetics and semiotics. Chance and/or necessity cannot program. Mass/energy cannot program. Physical constraints cannot generate formal controls. Only "choice with intent" at bona fide decision nodes can program.

No Ph.D. thesis was ever improved by typographical errors. Variation must be steered towards utility to be of any benefit. Otherwise, variation only corrupts existing Prescriptive Information (PI). This is why we go to such great trouble to protect messages in the Shannon channel from "variation" (noise pollution). It is also why life has such exquisite mechanisms of error prevention (e.g., Hamming redundancy block coding used in protein prescription;[2,5] bit parity[276,277,307] and error correction.[308-312]) Mere variation is the enemy of life's PI, not its benefactor.

Most "programming improvements" attributed to microevolution are in fact switched on/switched off controls of existing PI. The programing of these switches was already there, just inactive until switched on. Most bacterial resistance to antibiotics is a classic example. Under no circumstances could bacteria, even with a generation time of 20 minutes, evolve new prescriptions of function (resistances to so many different antibiotics) in just two to three days of exposure to a new antibiotic. That's not even sound evolution theory. The PI needed to adjust to antibiotic challenges was already there, in dormancy, waiting for environmental challenge and the need to be activated using existing configurable switches. The extent of pre-programmed sophistication is just all the more impressive.

What an amazing, powerful theory is this "duplication plus variation"! With a model like that, we can explain anything and everything—or absolutely nothing—as the case may be. Ask any computer programmer how far she would get using "duplication plus variation" to write new computational programs. And this is supposed to be biology's most confirmed theory, even proven fact, for explaining the spontaneous programming of both original and totally different new kinds of life?

Each and every more sophisticated genus of life from protocell to *Homo sapiens* was supposedly programmed by duplication plus noise pollution of duplicated PI. Statistical prohibitiveness is not kind to many scientific theories. Nevertheless, science tries to keep an open mind to extremely low probabilities. But, as we saw in section 3.20, possibility is not the same as plausibility.[330] Scientific method provides a means of definitive falsification of any theory with a Universal Plausibility Metric $\xi < 1.0$.[330]

Mutations result in loss of PI, not increased PI. Philosophic naturalism would have us believe otherwise. But the evidence is sorely lacking.

Biosemiotic entropy involves the deterioration of biological sign systems. The genome is a coded sign system that is connected to phenotypic outputs through the interpretive functions of the tRNA/ribosome machinery. This symbolic sign system (semiosis) at the core of all biology has been termed 'biosemiosis'. Layers of biosemiosis and cellular information management are analogous in varying degrees to the semiotics of computer programming, spoken, and written human languages. Biosemiotic entropy—an error or deviation from a healthy state—results from errors in copying functional information (mutations) and errors in the appropriate context or quantity of gene expression (epigenetic imbalance). The concept of biosemiotic entropy is a deeply imbedded assumption in the study of cancer biology. Cells have a homeostatic, preprogrammed, ideal or healthy states that are rooted in genomics, strictly orchestrated by epigenetic regulation, and maintained by DNA repair mechanisms. Cancer is an eminent illustration of biosemiotic entropy, in which the corrosion of genetic information via substitutions, deletions, insertions, fusions, and aberrant regulation results in malignant phenotypes. However, little attention has been given to explicitly outlining the paradigm of biosemiotic entropy in the context of cancer. Herein we distill semiotic theory (from the familiar and well understood spheres of human language and computer code) to draw analogies useful for understanding the operation of biological semiosis at the genetic level. We propose that the myriad checkpoints, error correcting mechanisms, and immunities are all systems whose primary role is to defend against the constant pressure of biosemiotic entropy, which malignancy must shut down in order to achieve advanced stages. In lieu of the narrower tumor suppressor/oncogene model, characterization of oncogenesis into the biosemiotic framework of sign, index, or object entropy may allow for more effective explanatory hypotheses for cancer diagnosis, with consequence in improving profiling and bettering therapeutic outcomes.[379]

A mutation must be immediately phenotypically beneficial for natural selection to have any hope of preserving it. Virtually all potentially beneficial mutations are found in the neutral zone of Kimura's graphs. They are silent mutations. Even if sophisticated repair mechanisms somehow miss them, their contribution to phenotypic fitness is negligible. No basis exists for their selection. They are also so rare that no curve can be visualized in Kimura's positive neutral mutation zone.[244] No imagined foothill incline exists on the back side of "Mount Improbable."[363] Mike Gene finds no hard evidence of any simpler precursor of genetic code. The genetic code appears to have existed in its current form in every simple life form from the beginning of known life.[380]

Abundant evidence exists of genetic entropy leading only to the long-term deterioration of genomic and epigenomic instructions, never to the increase in efficaciousness of those instructions. Gryder et al. review considerable evidence in support of The Prescription Principle from molecular biology (See Section 2.8) in their discussion of "Biosemiotic entropy of the genome."[379] John Sanford's work (Cornell University) has demonstrated a relentless general deterioration of the human genome.[381-383] The same deterioration appears to be going on through time in the genome of every other species. Population geneticists have known for a long time that no existing species will survive indefinitely. A highly deleterious mutation load builds up. Neutral mutations are non-selectable, and seemingly inconsequential. Neutral mutations are only *potentially* beneficial at the very best. Evolution cannot Select *FOR* potentially beneficial mutations. It only favors the best already-programmed, already-living phenotypic organisms.

DNA's repair mechanisms are constantly at war with mutations. As extremely rare as potentially beneficial mutations are, DNA repair mechanisms make their retention all the more unlikely.

3.25 Emergence

Emergence describes a supposedly spontaneous rise in integrative utilitarian controls and formal functionality of biosystems. The contention is, "Organized bio-function didn't have to be prescribed. It just emerged," meaning, "it just happened." Often large periods of time are added to the mix to make the absurd notion sound more palatable. "Emergence" is what philosophic naturalism appeals to in an attempt to justify belief in the spontaneous appearance of "information technology" and other types of engineered utility.

"Emergence" is a wonderfully scientific-sounding term for a completely vacuous concept with zero empirical support. It also offers no possibility of falsification. This latter point alone eliminates the notion right off from the realm of scientific investigation.

Included in the emergence fabrication is the dream of self-organization. Even *the term* "self-organization" is nonsensical. Nothing can self-organize itself into existence. This includes ribosomes that manufacture other protein molecular machines. Humans can only use the intelligence they find within their already existing brains and minds. It is simply a logical impossibility for *anything* to self-organize itself into existence.

Organization is an effect. It must be caused. Since organization is fundamentally formal, it can only be caused by formalistic causation, not by physical "cause and effect."

The first topic of "emergence" discussion should be to ask HOW, by what mechanism, did organized bio-function just "emerge?" What is the scientific basis for this notion? How was it established that faith in "emergence" is anything more than blind-belief superstition? Are we sure that the notion of emergence is not just a product of having to remain logically consistent with a purely metaphysical, materialistic, philosophic, commitment?

Included in belief in "emergence" is the spontaneous occurrence of prescription and its processing of non-trivial functionality. Computation and algorithmic optimization are just presupposed to occur naturally, from chance and necessity alone. Asking any scientific questions about this pre-assumption is invariably labeled "unscientific." It is simply not allowed.

Could inanimate nature have generated the "information technology" needed to craft life? Has any investigator ever observed a deterministic conglomerate of atoms or molecules, constrained only by initial physical conditions plus fixed natural law, spontaneously give rise to a "technology" of any kind? Note that we are not talking, at the moment, about life origin.

We are only talking about the "spontaneous generation" of the simplest "information-mediated technology."

Perhaps the most well-developed, comprehensive Protometabolism-First model published by any one author is Tibor Gánti's "chemoton theory."[384-390] Tibor Gánti's chemoton model is a hypercycle take-off (See section 3.25.1 below on Hypercylces). Gánti envisioned interconnected autocatalytic, or at least mutually catalytic,[391] Eigen-Schuster type hypercycles[133,392-397] in well-organized elementary units of life called "chemotons."[384,387-389]

Gánti's basic idea is that "stoichiometric cycles" act as catalysts. No proteins existed yet in his model, and therefore he proposes that no linear digital prescription was needed. He uses a cyclic process sign to replace equality in mass-balance equations.[62] The cycles supposedly do the work of enzymes and ribozymes. It's a bottom-up theory that is very appealing to those faithfully committed to purely materialistic metaphysical presuppositions. The model depicts a transition directly from chemicals to living "systems," without any formal controls.

Chemotons supposedly have three self-producing (autocatalytic) stoichiometric subsystems that are coupled to one another: autocatalytic metabolism, a spontaneously formed genetic polymer, and a membrane. The key to life is "fluid automata," complex systems of chemical reactions in fluid phase that function like machines. They have no solid parts. But they can be regulated.

Gánti's model is quite different from Lancet, Segre and Shenhav's composomes.[398-404] Gánti recognizes, and even emphasizes, the need for cybernetic controls. "Chemical reactions as building blocks can be assembled into regulated and program-controlled chemical automata without including any solid components."[390, pg xiv]

Gánti realizes that true organization is essential for life to exist. The chemoton model's "organizational principles must be present in every living being." Gánti calls his chemical cycles and networks "cycle stoichiometry." Gánti sees a direct link between genetic cybernetics and computer science. "Program control must control a functional system and enzymatic regulation must also regulate a functional system."[390, pg xii] "Chemoton theory is concerned primarily with this machinery aspect."

One can easily identify a major self-contradiction in this chemoton model. No source of bona fide controls is hypothesized, yet the need for controls is readily acknowledged. Understanding of the fundamental difference between physicodynamic constraints and formal controls is missing. Controls simply cannot arise without Choice Determinism (CD).

Says Gánti, "A chemoton is the simplest chemical machine which shows the generally accepted characteristics of life."[390] The first problem with this is that no such machine exists. The simplest living organism is a cell containing multiple operating systems, hundreds if not thousands of programs, multiple layers and dimensions of Prescriptive Information (PI), processing machinery, and huge numbers of molecular nano-computers all cooperating in one concerted integrative effort.[26,347]

The second problem is that "the generally accepted characteristics of life" depend upon who is defining life. No two scientists' definitions of life seem to be the same. Life-origin science has a long history of defining-down life to something far less than life in order to make our naturalistic models of life-origin "work for us."[405]

Gánti repeatedly refers to his chemotons as not only "interconnected systems of chemical reactions," but as "organized regulated processes." "Organizational principles must be present in every living being."[390, pg 1] The question of exactly how chance and/or necessity could sense, obey or pursue a formal "organizational principle" is never addressed. Until this missing essential piece of the puzzle is supplied, the model falls apart as a supposedly naturalistic explanation.

Gánti envisioned his fluid automata chemotons to have two parts: an operating part—the automaton; and a controlling part—the genetic programs.[390, pg. 13] Regarding the controlling part, Gánti makes a stunning admission: "Of course, *the sequence of signs is not material* [italics mine]. But neither is it independent of material, since the sign is carried by some material substance."[390, pg 13] In other words, Gánti regards the information found in the sequence of signs as nonmaterial even though that sequence is instantiated into a physical medium of monomeric sequence. It is surprising to this author that much of the supposedly controversial material presented in *The First Gene*,[5] Gánti himself would probably have had to agree with.

What in nature for Gánti could possibly be "not material" (not mass/energy)? Naturalistic science is physicalistic. Reality tends to be defined solely in terms of mass/energy. When Gánti talks about non-material representational sign syntax, what exactly is he talking about that could be encompassed by naturalistic presuppositions? Gánti would probably assure us of his commitment to a naturalistic worldview. The whole point of naturalistic life-origin science is to avoid nonphysical explanations of what is claimed to be a purely physical reality. But is his cybernetic model tenable under the naturalistic metaphysical imperative? Representationalism is a little hard to explain from a physicalist perspective! So are the mathematics and reason upon which the scientific method relies. To argue logically for physicalism is to deny physicalism. There is nothing physical about the exercise of logic theory in the defense of physicalism.

What does Gánti's word "program" mean? Programming requires purposeful logic-gate settings. Exactly how are programs instantiated into this liquid chemoton's physicality? What is the basic unit of selection and instruction in this liquid? By what chemical mechanism did this programming arise? Can we cite any examples of chemical reactions in liquid phase, especially, programming and optimizing algorithms? Says

Gánti, "The living system is a program-controlled cybernetic system."[390, pg 12] Gánti continues, "Cybernetics itself originated from the study of the regulated and controlled operation of living systems, and program control is already familiar from the genetic program."[390 pg 12] Notice how "program control" was just presupposed, but never explained, from within naturalistic metaphysics. Yet, we are expected to believe that naturalistic metaphysics is "scientific fact."

Says Gánti, "Chemical reactions as building blocks can be assembled into regulated and program-controlled chemical automata without including any solid components."[390, pg xiv] But what is the nature of this mysterious liquid program? Is it a liquid crystal? How could programmed information be instantiated into a liquid OR a highly ordered solid structure? And what exactly *does* the "regulating" of these chemical reactions? What sort of magic cybernetic liquid crystal is this? How did formal controls arise from chance and necessity, mass and energy?

There is no basis for logic theory, quantification, decision theory, scientific debate, computation, computer science, controlled experimentation, or engineering within the physicalist world view. But we ought not be too hard on Gánti. He at least acknowledges the reality of a representational sign system in the cybernetics of life. The sequence of commands physically instantiated into "physical symbol vehicle" (token) sequencing is the most important aspect of life. It constitutes the message of messenger molecules. But messages are formal, not physical, even though physical tokens are used to convey those messages.

The most serious problem with Gánti's model is that he is unable to generate a basis for "programming" OR "organization" from mere physicodynamics. Interconnected hypercycles are even more constrained than individual cycles. Constraints are not controls. Constraints cannot steer events toward pragmatic goals. We will examine in greater detail the problems with hypercycle theory below in Section 3.25.1.

Gánti's chemotons are assumed to be already evolved prior to the appearance of catalytic RNA's.[390, pg vii] So RNA linear digital prescription cannot explain Gánti's acknowledged need for programming and organization in the forming of chemotons. RNA was supposedly assembled only later by substrates already present in the chemoton.[390, pg vii] This notion alone does not explain how the particular functional sequencing of "signs" [actually tokens—"physical symbol vehicles" in a Material Symbol System (MSS)[44,45]] was achieved.

The spontaneous self-organization of ever-improving hypercylces, stoichiometric self-assemblies, and Gánti's chemotons[390] have never been observed, let alone repeatedly observed. No prediction fulfillments have ever been realized. "Self-organization" provides no mechanism and offers no detailed verifiable explanatory power. The hypotheses of chemotons' ever-growing capabilities are not even falsifiable. No lack of evidence, or the repeated observation of hypercycles' failure to arise, is capable of providing falsification. So the notion is conveniently and indefinitely protected from any scientific challenge. It must just be accepted by blind faith. Any scientist who raises an eyebrow of healthy scientific skepticism is immediately labeled a heretic by the hierarchy of scientism's presupposed imperative of metaphysical naturalism.

Little needs to be said about emergence, other than the fact that it simply doesn't occur. The only "emergence" ever observed is the self-ordering phenomena of chaos theory. But self-ordering has nothing whatever to do with organization or functionality, and certainly nothing to do with metabolism or life (See Section 3.11).[62,81]

Maybe Venter, Dawkins, Church, Orgel and Johnson and Gánti are all wrong about the need for organization, prescriptive information, regulation and controls. Has anyone been able to discount their perspective by presenting a life form that exists independent of information technology? Literally, by the day, new peer-reviewed science journal papers are published worldwide confirming all the more that *information processing is the key to life*. And this information technology not only predates Homo sapiens, it predates vertebrates and even metazoans (multi-celled organisms). In fact, it predates eubacteria and archaea. It predates LUCA (the Last Universal Common Ancestor), the Most Recent Common Ancestor (MRCA) of all current life on Earth, and the very first cell. Even a protocell could not organize without the steering of large numbers of biochemical pathways toward protometabolic success. Prescriptive Information (PI)[2,21,22,31,42,82] was needed to instruct and materialize that organization. Algorithmic processing of those instructions was also needed,[23,24,26-28,315,355] involving hardware, firmware, wetware, operating systems, and additional software. The information technology of the simplest known cell puts to shame the most sophisticated artificial technology ever designed and engineered by the most brilliant of human minds.

Walker and Davies, in a paper entitled "The algorithmic origins of life," argue,

> Although it has been notoriously difficult to pin down precisely what is it that makes life so distinctive and remarkable, there is general agreement that its informational aspect is one key property, perhaps the key property. The unique informational narrative of living systems suggests that life may be characterized by context-dependent causal influences, and, in particular, that top-down (or downward) causation—where higher levels influence and constrain the dynamics of lower levels in organizational hierarchies—may be a major contributor to the hierarchal structure of living systems.[361]

Walker and Davies go on to discuss the onset of the unique mode of (algorithmic) information processing characteristic of living systems.

A popular review of this paper states:

Scientists trying to unravel the mystery of life's origins have been looking at it the wrong way, a new study argues. Instead of trying to recreate the chemical building blocks that gave rise to life 3.7 billion years ago, scientists should use key differences in the way that living creatures store and process information, suggests new research detailed today (Dec. 11) in the Journal of the Royal Society Interface.

Study co-author Paul Davies, a theoretical physicist and astrobiologist at Arizona State University says:

> In trying to explain how life came to exist, people have been fixated on a problem of chemistry, that bringing life into being is like baking a cake that we have a set of ingredients and instructions to follow. That approach is failing to capture the essence of what life is about.
>
> Living systems are uniquely characterized by two-way flows of information, both from the bottom up and the top down in terms of complexity, scientists write in the article. For instance, bottom up would move from molecules to cells to whole creatures, while top down would flow the opposite way. The new perspective on life may reframe the way that scientists try to uncover the origin of life hunt for strange new life forms on other planets.[406]

Davies was certainly correct about biology's fixation on "chemistry." Recipe "instructions" may not be all that bad a description of much of the uniqueness of Life. All cells depend uniquely upon formal "instructions" and the algorithmic processing of those instructions. The prescriptive informational and computational realities of life are what have been ignored for far too long.

Agnostic NYU Philosophy Professor Thomas Nagel recently jolted the world with a new book and journal paper entitled, *Mind and Cosmos: Why the Materialist Neo-Darwinian Conception of Nature Is Almost Certainly False.*[407,408]

> It is no longer legitimate simply to imagine a sequence of gradually evolving phenotypes, as if their appearance through mutations in the DNA were unproblematic . . .[407, pg. 9]

> Doubts about the reductionist account of life go against the dominant scientific consensus, but that consensus faces problems of probability that I believe are not taken seriously enough, both with respect to the evolution of life forms through accidental mutation and natural selection and with respect to the formation from dead matter of physical systems capable of such evolution.[407, pg. 9]

> The coming into existence of the genetic code -- an arbitrary mapping of nucleotide sequences into amino acids, together with mechanisms that can read the code and carry out its instructions -- seems particularly resistant to being revealed as probable given physical law alone.[407, pg. 10]

Other papers focus on DNA's information theory and storage capacity capabilities.[308,362,409]

> What can be said about 'emergence,' except that it is nothing less than outright superstition. No one has ever observed it. Like 'teleonomy,' it is a manufactured hypothetical naturalistic term for a nonexistent phenomenon. Physicality alone cannot generate computational success and non-trivial bio-function. Mass and energy cannot steer physicodynamics towards sophisticated utility.

3.25.1 Hypercycles

Purely physicalistic hypercycle models[133,393-395,397] consist of nothing more than self-ordered circular constraints, not organizational controls. No selective steering toward function or formal work has ever been demonstrated to arise spontaneously out of chance or physicodynamic determinism, hypercycles included. Hypercycles, if they exist at all, are a form of self-ordering, not organization.[62] Physicodynamic constraints are not formal controls.[61] They provide no formal controls needed for life's information technology (See section 3.23).

Even if hypercycles did exist, they would only consume all available resources in a redundant, unimaginative, monotonous spiral. Circular constraints do not actively, or even passively, climb foothills of rugged landscapes toward mountain peaks of optimized formal function. Eigen's hypercycles are nothing more than a pipe dream when it comes to the spontaneous generation of formal organization and utility. This is why we have never seen a single such phenomenon in the real world since all of Eigen's publications on hypercycles in the 70's and 80's. We always awake from our naturalistic dream back into the disappointing reality of "No new non-trivial prescription arises from chance and necessity." Any experimental support for the notion of hypercycles always involves hidden experimenter steering ("investigator involvement" in experimental design), usually through artificial selection at successive iterations. That is *not* evolution.

The inanimate environment could care less whether anything functions, let alone functions optimally. No perception of mountain peaks of formal utility exist in a prebiotic environment. An inanimate environment doesn't even perceive foothills of formal function. Whatever happens physicochemically, happens. There's no "good' or "bad" with reference to utility in a prebiotic environment. Inanimate nature is blind to function.

Any naturalistic model that presupposes inanimate nature's preference for superior function is totally without empirical and sound experimental support. Strip Materials and Methods of its hidden formal controls, and the experiment produces nothing but useless tar. Spontaneous hypercycles of ever-increasing functional sophistication simply do not occur.

3.25.2 Trial and Error Searches

Section 3.17 showed that trial and error searches, no matter how inefficient, are nonetheless still *pursuits*. "Trial and Error" searches are iterative. They require at least a "Yes" or "No" answer to each question in a series of questions that ideally keep halving the remaining possibilities. Even when the remaining possibilities are halved with each successive question, this kind of search is still inefficient. But, for each question to half the remaining possibilities, those questions must be incredibly conceptual and highly educated in their phrasing.

Most important of all is that even the most inefficient sequence of questions is still a teleological process. Changing the label to "teleonomic" does nothing to change the teleological nature of the search for a goal or target. Any such search is simply not "natural."

Trying to turn evolution into a programming modality by calling the pursuit of function "trial and error" won't fly. Declaration of value, desire for utility, and pursuit of goal are all formalisms inherent in "trial and error" searches, the same as with any other kind of search. Inanimate nature does not pursue any goal.

No imagined hypercycle could possibly conduct a trial-and-error pursuit of formal function, let alone organize and program life.

3.25.3 Oscillation Models of life origin

What empirical evidence do we have that mere oscillation around bifurcation points can produce strings of wise steering and programming choices at true decision nodes? Mere oscillation is like coin-flips at decision nodes. Coin flips render "decision nodes" nothing more than "bifurcation points." Bifurcation points are nothing more than forks in the road. Forks in the road represent "contingency" and "possibility." Multiple pathways are possible. But possibility does not measure up to probability, and probability does not measure up to plausibility. Neither probability nor plausibility add up to computationally halting programming, or metabolically productive prescription of cooperative binding functions. Mere oscillation at bifurcation points does not prescribe metabolic pathways, cycles or metabolic schemes.

Mere bifurcation points tell us nothing about *which path should be taken* to shorten our journey. Mere bifurcation points tell us nothing about which choices achieve utility or optimize functionality. Only insightful, purposeful, programming choices produce eventual sophisticated bio-function and successful computation.

The chief oscillation proponent is Vladimir Kompanichenko.[410-414] Kompanichenko envisions non-equilibrium transitions - from initial to new stable states, through mere bifurcation points. No empirical justification exists to just assume that the new stable state has any functionality or utility. Such expectation is just hoped for. "Relative stabilization" is close to equilibrium, not a state "far from equilibrium." The only thing that moves physicodynamics opposite from an equilibrium tendency is purposeful programming choices and formal organization. The latter requires CCCC, or CD.

The oscillation model appeals to imagined synergetics:

1) Instability is generated in the temporal bifurcate position (between the stable states).

2) At a bifurcation point, accidental changes occur in the system. These influence the path of its further development, producing uncertainty in the path choice and a new stable state.

3) In general, the system after the bifurcation point may either ascend to higher order/organization, or descend to lesser order/organization.

The contention is made, "if a system oscillates around a bifurcation point (BP), entropy will rise after that BP (descending trend), and also spontaneously decrease (ascending trend)." This would be a direct violation

Section 3: How did primordial prescription, and its processing, arise in a prebiotic environment?

of the Second Law without some specific exception being enumerated. The only known exception comes via formal intervention. Only wise programming and purposeful choices are able to locally and temporarily circumvent the Second Law, never mere physicodynamics alone. Physicodynamic events themselves always obey the Second Law. Even chaos theory obeys the Second Law. The only temporary and local circumvention of the Second Law comes from prescribed steering toward formal pragmatism.

In the oscillation model, internal supposedly "physical information" is believed to fluctuate, decreasing and increasing. But this model misdefines "information." Combinatorial uncertainty can spontaneously increase in a purely physical world, but not Prescriptive Information (PI). Prescriptive information can only arise through formal choice-contingency, not through mere physical interactions like phase changes and chemical reactions. The latter are blind to the pursuit of organization and potential function.

The notion that mere oscillation can generate sophisticated organization and function also makes the mistake of thinking that PI is physical. PI is formal, not physical, just like mathematics, symbol systems, language, logic theory, and cybernetics are nonphysical and formal. But, of course, we can record purposeful choices into physical media using a formal material symbol system (MSS). We can use molecules of ink and paper, electrons, email, Morse code impulses, or smoke signals to send meaningful, functional messages. But the molecules or electrons themselves cannot create meaning or sophisticated function.

Belief in emergence from mere oscillation at bifurcation points requires faith that the thermodynamic inversion "information" begins to accumulate faster than it dissipates, forming biological information. Zero empirical evidence exists of this ever having happened. Without purposeful programming choices being made at bona fide DECISION nodes, no organization or sophisticated function has ever been observed to just "emerge." Mere bifurcation points are not decision nodes. They are simply choice *opportunities*. But if no agency is available to take advantage of choice opportunity, no sophisticated function arises.[339]

Presupposed without evidence is a spontaneous increase in formal function. Emergence is just blindly believed. Uncertainty is envisioned to achieve what only efficacious choices produce. In the absence of wise programming decisions, the new "stable state" is unrelated to formal "usefulness" for any purpose, and will always manifest increased entropy.

Second, the envisioned new state is not a new *stable* state. Dissipative structures are dissipative! That means momentary, not sustained. Dissipative structures are anything but stable. The only reason a tornado looks "sustained" is a long succession of momentary dissipative states, creating the illusion of a stable structure.

Very sophisticated Sustained Functional Systems (SFSs, e.g., machines and engines) are needed to sidestep entropy increase.[30] Mere Prigogine "dissipative structures" are not sustained. This is the very reason they were called "dissipative" by Prigogine.[288] A tornado is actually a string of rapid-succession dissipative structures. But even that string is momentary, and dissipates as rapidly as the tornado self-ordered.

An increasing number of well-known, responsible scientists are beginning to question just how scientific is faith is in the spontaneous self-organization of even a protometabolism, let alone metabolism. Spontaneous increase in higher ORDER has been well-substantiated by Prigogine and many successors, but *not* higher ORGANIZATION! (See Section 3.11)

The only thing that really decreases entropy and improves organization is formal choices made with intent and in pursuit of sophisticated function. But inanimate nature cannot make choices with intent, and does not pursue any such goal. Not even evolution has any goal.

Formal choices can be instantiated into physical objects (e.g. tools) and processes (e.g., manufacturing plants), but not without goal and formal agency.

No scientific fact exists in support of mere oscillation at bifurcation points generating programming instructions (genomic or epigenomic). Oscillation could not have produced formal organization and non-trivial function, even on the primordial level of a protocell.

3.26 Definition of life

All attempts to define life have failed. We are forced to settle for descriptions of life rather than a definition of life.[127,415,416]

An international conference of university professors was held in Modena, Italy at the turn of the millennium.[188,417] The purpose of the conference was to define life. Every participant was required to submit, in advance, a definition of life. Out of a couple hundred life-origin investigators participating, no two definitions of life were the same.

No other single factor contributes more to an adequate description of life than the factor of tight control and/or regulation. If the definition of life could be reduced to just one word, it should probably be "Control." The inanimate world is ruled by constraints. Yes, life is also subject to constraints, but life is uniquely ruled and made possible only by formal controls.[19,21,22,29,31,42,82]

The Prescriptive Information (PI)[2,5] in the cell provides instructions and processing for that control, usually in the form of genomic programming and epigenetic configurable switch-settings that integrate and switch on and off cellular circuits. Alternative splicing in cells serves as the equivalent of higher level programming in man-made cybernetic machines. The editing of mRNA, along with the many newly discovered layers and dimensions of intracellular PI are mind-boggling. The simplest of cells manifests higher programming languages and mind-boggling corrective controls, regulation and algorithmic processing.

Genetics is not the only source of control in cells. Genetics has been largely antiquated by genomics. Genomics is still in the process of being antiquated by epigenomics. Configurable switches are turned on and off in nucleotides, for example, that determine gene activity. This constitutes epigenetic control mentioned above. Even though the sequence of nitrogen bases in a gene remains the same, the cell's metabolism is controlled and regulated by other factors such as methylation and acetylation programming of true *logic gates*. We do not see such formal phenomena in inanimate nature. Cybernetics is unique to life. Cybernetics requires wise purposeful choices at true decision nodes. Chance and necessity cannot control. They can only constrain, independent of any functional goal.

Controls are everywhere active in the simplest known living organisms (e.g., *Mycoplasma genitalium*). Controls are completely absent in inanimate physics and chemistry. Constraints and fixed laws cannot produce computational halting and metabolic success.

Another unique attribute of life is use of a representational symbol system where the physical symbol vehicles (tokens) serve in a more formal capacity than just physicochemical reactants. The nucleotides do not chemically react with amino acids, for example. A triplet codon of nucleotides *formally represents and prescribes* each amino acid or stop code. Prescriptive causation is found in an abstract, conceptual, formal codon table, not in direct binding or physical catalysis.

Below is a descriptive list of invariant characteristics of free-living life. Sustained, free-living life is any system which from its own inherent set of biological instructions and algorithmic processing of that Prescriptive Information (PI) can perform all nine of the following bio-functions (Used with permission from http://www.lifeorigin.info):

Section 3: How did primordial prescription, and its processing, arise in a prebiotic environment?

(1) Delineate itself from its environment through the production and maintenance of a membrane equivalent. In theoretical early life, this membrane equivalent would most likely have been a rudimentary or quasi-active-transport membrane necessary for selective absorption of nutrients, excretion of wastes, and overcoming osmotic and toxic gradients,

(2) Write, store, and pass along into progeny Prescriptive Information (PI) (instructions, both genetic and epigenetic / epigenomic) needed for organization; provide instructions for energy derivation and for needed metabolite production and function; symbolically encode and communicate functional message through a transmission channel to a receiver / decoder / destination / effector mechanism; integrate past, present and future time into its biological PI (instruction) content (PI instructions can be implemented now or any time in the future. In addition, according to evolution theory, these instructions embody a protracted history of derivation and former control. PI is thus largely time-independent, a feature that bespeaks its formal rather than physical essence.),

(3) Bring to pass the above recipe instructions into the production or acquisition of actual catalysts, coenzymes, cofactors, *etc.*; physically orchestrate the biochemical processes/pathways of metabolic reality; manufacture and maintain physical cellular architecture; establish and operate a semiotic system using signal molecules,

(4) Capture, transduce, store, and call up energy for utilization (intuitive, useful work, not just the physics definition of "work"),

(5) Actively self-replicate and eventually reproduce, not just passively polymerize or crystallize; pass along the apparatus and know-how for homeostatic metabolism and reproduction into progeny,

(6) Self-monitor and repair its constantly deteriorating physical matrix of bio instruction retention/transmission, and of architecture,

(7) Develop and grow from immaturity to reproductive maturity,

(8) Productively react to environmental stimuli. Respond in an efficacious manner that is supportive of survival, development, growth, and reproduction, and

(9) Possess enough relative genetic stability, yet sufficient mutability and diversity, to allow for adaptation and potential micro evolution.

All free-living classes of archaea, eubacteria, and eukaryotes meet *all* nine of the above criteria. Eliminate any one of the above nine requirements, and it remains to be demonstrated whether that system could remain alive. RNA strands, DNA strands, prions, viroids, and viruses are not free-living organisms. They fail to meet many of the above well-recognized characteristics of independent life.

Even in historical and theoretical science, there must be some degree of empirical accountability to our theories. Models of life origin *must not consist of defining down the meaning and essence of the observable phenomenon of life to include nonlife in order to make our theories work for us*. Any scientific life-origins theory must connect with life as we observe it (the Continuity Principle). Science will never be able to abandon its empirical roots in favor of purely theoretical conjecture.

On the other hand, science must constantly guard itself against Kuhnian paradigm ruts. We must be open-minded to the possibility that life has not always taken the form that we currently observe. In addition, we must take into consideration the problems inherent in any historical science where the observation of past realities is impossible.

We could propose a population of one-cell look-alikes consisting of exceedingly naïve pseudo membranes encircling stochastic ensembles of an RNA analog. But we would have to really stretch our imagination to the point of Freudian wish–fulfillment to legitimately consider such "look-alikes" as "protocells," let alone "primordial cells." Because a soap bubble "buds" into two soap bubbles, for purely physicodynamic reasons, does not make that soap bubble a yeast cell. To think that such one-cell look-alikes could progress into supposedly living protocells without programming is ludicrous. All evolutionary theory rests squarely upon "selection." But few evolutionary theorists appreciate that all they have to work with in their model is Selection *FROM AMONG*. Macroevolution is impossible from Selection *FROM AMONG* alone. The programming of any non-trivial increase in sophisticated PI requires Selection For (in pursuit of) at true decision nodes. Evolution is only eliminative.

3.27 Challenges remaining in astrobiological research

Astrobiologists are very accustomed to addressing an array of difficult problems extending well beyond astronomical mysteries. Included are homochirality and a long list of chicken-and-egg biochemical paradoxes. Far more perplexing, however, are the remaining challenges relating to the initial formal organization and control of primordial cells.

Sophisticated programming could not have been random or law-induced. Each nucleotide in the string had to be selected with programming intent. Every nucleotide selection represents a two-bit quaternary (four-way) decision node. Each choice often affects multiple layers or dimensions of programming. Nucleotide selection could only have been selected by one of the two classes of selection in the Universal Selection Dichotomy (USD)—by Selection *FOR (in pursuit of)*, not Selection *FROM AMONG (existing already-programmed, already-living phenotypes)*. Even the back side of "Mount Improbable" is way too steep for inanimate nature to have programmed all of these choices.

Natural selection consists only of Selection *FROM AMONG,* never Selection *FOR (in pursuit of).* Yet the programming of life undeniably consists of Selection *FOR (in pursuit of)* at bona fide decision nodes. Natural selection cannot possibly have organized and programmed life. No selection "pressure" existed until *after* already-programmed, already-living phenotypic organisms could differentially survive and reproduce. Evolution is thus a non-theory when it comes to abiogenesis.

Mutations occur at the genetic level. Environmental selection, however, works only at the living organismal, phenotypic level. Neither physicochemical forces nor environmental selection chose the next nucleotide to be added to the prescriptive messenger biopolymer. Functional nucleotide choices are all polymerized by the same 3'5'phosphodiester bonds. We cannot attribute innovative, formal, metabolic schemes to chemical determinism.

By what mechanism did *prebiotic* nature select functional nucleotide sequencing? In the initial primary, positive, single-strand template, base pairing is not a factor in determining sequence "choices." Each nucleotide selection constitutes an initial algorithmic switch-setting. The syntax and circuitry of these switch-settings is what programmed the first few RNA analog ribozymes. How was RNA folding-function anticipated when covalently-bound primary structure was forming? Were these selections random, or constrained? If our naturalistic response is that there was no anticipation, only happenstance, what is the probability of so many hundreds of nucleosides per ribozyme, plus so many hundreds of needed three-dimensional ribozymes occurring at the same place and time? How plausible is such a model?

Could non-ribosomal short peptides have formed? Yes. That might have provided some additional crude catalysis. But very little. Rarely do more than ten amino acids polymerize spontaneously in an aqueous environment. Hydrolysis counters dehydration synthesis. Even if sophisticated ribosomes could have formed, the likelihood of a single polypeptide with *any* functional fold forming spontaneously is only one in 10^{77}.[418] It doesn't take a very long string before random generation of a needed primary structure becomes statistically prohibitive, and utterly implausible, as a proposed mechanism for prescription of sophisticated function.[80] And there are just too many hundreds of catalysts, each with the

Section 3: How did primordial prescription, and its processing, arise in a prebiotic environment?

exact required function, that were needed at the same place and time to have any hint of metabolism develop.

Were the selections made by some mysterious yet-to-be discovered law? If so, we would have had a lot of high probability polyadenosines, for example, with extremely low Shannon uncertainty, and close to zero bits of "possibility." Information retention would have been impossible if some yet-to-be-discovered law caused the sequencing.

Even within the most primitive of cells, no biochemical reaction seems unnecessary. Byproducts are recycled in highly imaginative ways. No energy expenditure seems wasted. Every physicochemical interaction contributes to the pursuit of a unified, coherent, holistic meta-metabolism. Every scheme is ingenious. Highly sophisticated molecular machines are hard at work around the clock with highly fine-tuned utilitarian performances. These sophisticated bio-functions cannot be reduced to chance and necessity, mass and energy. They are steered toward the goal of formal metabolic success. They are tightly controlled and regulated.

3.27.1 The quandary of life origin is not a matter of complexity; it is a matter of *conceptual* complexity.

Maximum complexity is nothing more than maximum Shannon uncertainty. Maximum uncertainty and complexity refer only to the greatest number of combinatorial possibilities. Complexity has nothing whatever to do with functionality. Stochastic ensembles (e.g., random strings of nucleotides or amino acid residues) are maximally complex. But they don't DO anything useful. In fact, only one out of every 10^{77} 100-mer polyamino acid strings fold into *any* bio-functional fold,[418,419] let alone a certain needed fold for a certain needed bio-function in a certain metabolic scheme. Complexity tells us absolutely nothing about utility, or lack thereof. Yet the word "complexity" has become the buzz word of naturalistic science's attempt to explain sophisticated algorithmic optimization in cellular metabolism. To try to explain computation by pointing to mere "complexity" is ludicrous.

What generates metabolic schemes, meta-metabolisms and life is *conceptual* complexity. What does "conceptual" mean? To conceive is not just to imagine or envision abstract functional possibilities. To conceive often includes actually conjuring up, designing, creating, inventing, engineering, manufacturing and bringing into existence, pragmatic theoretical ideas and plans. Robots, for example, are always conceived by artificial intelligence experts before being produced. They are "dreamed up," devised, formulated, and computed after first being abstractly imagined. The imagined goal alone is what steers the engineering to achieve the desired functional product.

Robots are not just complex. They are conceptually complex. Chance-contingency has never produced a robot. Neither has necessity (fixed, forced law-like regularity of physical interactions). Only one thing produces robots—concept. Of course, the nonphysical, abstract concept has to be instantiated into physicality before it can accomplish useful work in a physical realm. But physicality (chance and necessity, mass and energy) has never produced a single automaton.

Figure 6, this Section 3.27.1, shows the disconnect between mere complexity vs. *conceptual* complexity.

Section 3: How did primordial prescription, and its processing, arise in a prebiotic environment?

Figure 6. Complexity is often confused with programming controls and formal organization. The degree of three-dimensional structural complexity within a pile of pick-up sticks is staggering. But what exactly does this enormous degree of structural complexity DO? If we poured glue on this pile to freeze its structure, what sophisticated formal function would this *complex* pile of objects generate? Mere combinatorial complexity must never be confused with organization or formal utility. (Used with permission from Abel DL: The capabilities of chaos and complexity. *Int J Mol Sci* 2009, 10:247-291.)

Life is the most *conceptually* complex phenomenon known to science, far exceeding any man-made invention in its organization, operational design, algorithmic optimization, and engineering. To refer to life with nothing more than a term defining "many possible alternative combinations" (complexity) is laughable. Life is complex, to be sure. But it was not organized by mere complexity. It was organized only by concept in advance of its existence.[420] Life could not have come into existence without thousands of highly optimized algorithmic processes, all being up and running.[336] In addition, all these processes had to be holistically integrated to promote the goal of being and staying alive.

But, what exactly is concept? What is required in order to conceive? First, there must be freedom from necessity in order to imagine anything. Bona fide decision nodes must exist at which choice-contingency, not just chance-contingency, can be exercised. Motive is always behind conception. Choices at decision nodes must be purposeful in pursuit of pragmatic goals. Inanimate nature knows nothing of such goals and pursuits. It cannot choose from among options in pursuit of "usefulness." The environment is blind to usefulness at the programming stage. It may secondarily "prefer" the best already-living organisms through differential survival/death of already-living small groups of organisms. But the environment has no creative ability. *Evolution is only eliminative.*

Even the smallest known free-living bacterium (*Pelagibacter ubique*) has 1.3 Mb of binary choice place holders in its DNA. 1,354 open reading frames are involved in 1,389 total genes.[421-426] But, even P. Ubique still has intergenic spacers, non-coding RNA's and sophisticated epigenetic configurable switch controls that greatly compound its conceptual complexity.

A popular media article in 2012 was entitled, "To Model the Simplest Microbe in the World, You Need 128 Computers."[427] This article summarized a paper from the science journal *Cell*[428] in which the computational requirements needed to model Mycoplasma genitalium were reviewed. Keep in mind that *Mycoplasma genitalium* has a much smaller genome than the smallest free-living organism, Pelagibacter *ubique*. *Mycoplasma genitalium* is considered by most to be an obligate intracellular parasite. If 128 computers are needed to process genomic integration in an obligate intra-cellular parasite with only 525 genes, imagine the computational power needed to process 1,389 genes in *Pelagibacter ubique*! How was this simplest known free-living cell prescribed and processed into existence in a prebiotic environment ruled by nothing but chance and necessity (fixed law)? Chance and necessity cannot make purposeful programming choices at decision nodes and configurable switches (logic gates). Even a protocell one tenth as conceptually complex could not have been generated by chance and necessity alone.

3.27.2 Multiple layers and dimensions of Prescription are superimposed in genomes.

Abiogenesis models are already conceptually complex enough to dumbfound any self-honest philosophic naturalist. But then we must add to the mix many other ingenious controls that would have had to arise fairly early on:

- negative strand regulation of positive strands;
- alternate splicing of gene components from multiple chromosome sites,
- still other forms of mRNA editing,
- RNAi inhibition,
- cytosine methylation,
- histone acetylation
- other three-dimensional prescriptions of function,
- translational pausing's control of folding,[253,254]
- post-translational editing,
- required independent chaperones, many additional kinds of switching controls (not constraints) being discovered by the month and reported in the literature

Multiple layers of information exist, including chromatin three-dimensional information.[429] Genes are pieced together from distant locations.[430-434] Edited nucleotide syntax ultimately prescribes minimum Gibbs-free-energy sinks of translated polyamino acid strings.[435,436] D'Onofrio,[254] following up on the work of Li, Oh and Weissman,[253] has even shown that functional folding is largely controlled by purposeful Translational Pausing (TP) at the back end of the ribosome, which in turn is determined by PI_o hidden in the supposed "degeneracy" of the codon table.

Codon redundancy, as Yockey pointed out decades ago,[437-439] is anything but "degenerative." Genomic PI_o is multi-layered and multi-dimensional. Chaperone proteins,[440-443] which aid in steering functional folding,[280,281] are themselves prescribed by edited nucleotide sequencing. These in turn determine three-dimensional protein folding into highly tailored molecular machines.[444-449] Regulation is carefully controlled with multiple kinds of configurable switches such as cytosine methylation[450-452] and histone acetylation.[453-455] Such informational phenomena, along with many others, provide way too much *formal* instruction, organization and function for philosophic naturalism (chance and necessity) to explain. No explanation for the origin of this PI_o is found within "differential survival and reproduction of already-programmed, already-living phenotypes" (natural selection: NS). NS does not work at the molecular/genetic level. It works only on small populations of living phenotypic organisms (see The Genetic Selection [GS] Principle[68,71,83]). All of the more recent epigenetic and decentralized control findings only further emphasize the cybernetic nature of life and the reality of the highly integrated PI_o that directs it.

Not only initial organization, but also regulation requires programming instructions. Transcription and translation rates are both carefully controlled *in response to, not by,* environmental flux. Living systems are programmed to meet their functional needs in most any environment. The environment is not the programmer, and not the source, of Ontological Prescriptive Information (PI_o). The genomic and epigenomic programming predated all environmental eventualities, and stood ready and able to respond to almost *any* environmental stress or opportunity. This *prior anticipatory programming* is the essence of biological PI_o.

Biological messaging systems provide further proof of the reality and essential role of PI_o.[54-57,273,456,457] These messaging systems often consist of transient multi-protein complexes that assemble cooperatively.[458] Say Bolanos-Garcia et al,

> . . . such assemblies are unlikely to form by chance, thereby providing a sensitive regulation of cellular processes. Furthermore, selectivity and sensitivity may be achieved by the requirement for concerted folding and binding of previously unfolded components.[458]

Bolanos-Garcia et al continue,

> We show that multi-protein assemblies moderate the full range of functional complexity and diversity in the two signaling systems. Deciphering the nature of the interactions is central to understanding the mechanisms that control the flow of information in cell signaling and regulation.[458]

Ornate, large, extremophilic (OLE) RNAs represent a recently discovered non-coding class of RNAs found in extremophilic anaerobic bacteria.[459] Biopolymers and co-enzymes require careful steering to the right place, and critical timing of bonding, catalysis, and physicochemical interaction. Critical energy management is needed at every turn. Metabolic function depends upon precise *prescription* of coordinated function. We cannot deny the need for and the reality of biological *control*. Control depends upon instructions. Biological instructions are a form of PI_o.

Further confounding our understanding of bio information are still additional layers, dimensions, and interleaving of epigenetic and extra genetic controls.[460,461] DNA methylation and covalent histone modifications play major roles in growth and development.[462] Variable histone modifications have been found to be present at metastable epi-alleles. This suggests that DNA methylation acts in concert with histone modifications to affect inter-individual variation of metastable epi-allele expression.[463] These epigenomic controls, along with histone modifications and other decentralized/disbursed/distributed controls, are highly informational and clearly cybernetic in nature.

Micura, et al. found that "nucleobase methylations at the Watson-Crick base pairing site provide the potential not only to modulate but to substantially affect RNA structure by formation of different secondary structure motifs."[464] Some bio information is even apparent in the physical spatial arrangement of chromosomes.[462,465,466] Multifarious machine components interact at multiple processing levels. Formatting and indexing are not just combinatorially complex, but conceptually complex. Cells can reprogram their own DNA, with heritable phenotypic results.[463,467,468] 5' Cytosine methylations, in particular, can silence genes altogether, alter their expression, remodel chromatin, effect and affect imprinting.[469,470] Methylation seems to expand the DNA material symbol system into a base 5 rather than base 4 symbol system.

Assuming, for the moment, total disagreement with many of the above contentions, let us try our hand at meeting the challenges listed below from a purely physicalistic perspective.

1. Why would a prebiotic environment have valued or sought to preserve function over non-function?

2. Why would mere chemical interactions have preferred mountain peaks of functionality over valleys in rugged fitness landscapes? What exactly was the impetus for mountain climbing? Reaching hundreds of these mountain peaks at the same time and place would have been required *before* even protolife could begin.

3. How far would "metabolism first" models have progressed if they had to keep "re-inventing the wheel" with every protocell death? In the absence of heritable Prescriptive Information (PI) in a material symbol system, and its sophisticated machine processing, each new protocell would have had to start all over from scratch.

4. How did nucleosides get specifically sequenced into computationally successful programming strings (Turing tapes)? The chemical bonds between them are all the same. Sequencing is independent from the laws of physics and chemistry. Physicochemical determinism cannot

Section 3: How did primordial prescription, and its processing, arise in a prebiotic environment?

explain functional sequencing, which requires freedom from physical determinism. Carbon chemistry provides this freedom from law (contingency). But, chance-contingency cannot program sophisticated function. What force programmed the needed *cybernetic* sequencing of nucleotide and codon tokens?

5. How did a Turing machine just happen to arise in inanimate nature at the same time as a Turing tape? How did the needed ability to process those Turing tapes simultaneously arise?

6. If non-enzymatic templating laid the foundation for life, what functionally sequenced the *template*?

7. How did ribosomes get organized into such sophisticated machines prior to the existence of life? Any notion of protoribosomes won't fly. A certain minimal threshold of sophisticated ribosomal structure and function exists below in which no proteins could have been made.

    a. Dozens of specific sub questions could be asked about specific ribosome mechanics (e.g., translational pausing mechanisms critical to protein folding), and how each of these sophisticated machine functions was achieved by inanimate nature.

8. If ribosomes are essential to protein manufacture, where did the many proteins come from that are integral to the ribosome's own structure and function?

9. How could the sequencing of codons have "known" what sequencing of amino acids to prescribe? Reverse programming models of code origin, with far simpler codon tables, are still fraught with great conceptual implausibility. Great obstacles also exist to table expansion later in time, after life already existed.[471,472]

10. How did the codon table's *formal rules* get written? Rules are quite different from laws, just as controls are quite different from constraints. Formal rules and controls are arbitrary and nonphysical.

They are choice-contingent, not chance-contingent. Physical law cannot generate rules or controls. Neither can chance.

11. Correct folding had to be instructed for every needed protein molecular machine to be reliably manufactured. How could minimum-Gibbs-free-energy folding requirements of polyamino acid strings have been anticipated when prescriptive polynucleotide sequencing was polymerized in advance with strong 3'5' phosphodiester bonds?

12. Even if proteins could form directly from polypeptide polymerization, how would polypeptide primary structure (sequencing with rigid peptide bonds) have achieved the minimal Gibbs free-energy of R-group-sequencing, spatial distribution, charge, and hydrophobicity needed for proper folding to produce each specific molecular machine?

13. How did superimposed layers of Prescriptive Information (PI) arise spontaneously, such that each nucleotide symbol selected at each locus in the rigidly-bound strand contributes to completely different semiotic strings of instruction, often going in opposite directions?

14. How did complementary base pairs, which we presuppose are determined solely by physicodynamic cause-and-effect, come to prescribe completely different sets of formal instructions when DNA was separated into independent strands?

15. How did prebiotic nature generate abstract *editing* of mRNA strings? If our answer is that protocells did no alternative splicing, how could a protocell have come to life with such simplistic prescription (the old one gene-one protein model)? The number of genes needed right off would have greatly increased.

16. What correlated all of the specific amino acyl t-RNA synthetase correspondences with each amino acid?

Section 3: How did primordial prescription, and its processing, arise in a prebiotic environment?

17. What correlated each anti-codon with each specific tRNA with each specific amino acid?

18. How did nature know how to write Hamming-redundancy block codes to reduce noise pollution in the Shannon channel (poorly termed codon "degeneracy")? There is nothing "degenerate" about these block codes, which manage multiple layers and dimensions of prescription.

19. "Silent SNP's" (e.g., variations of the third nucleoside in the wobble position of a codon) are not silent. So-called "degenerate" codons are highly instructive of translational pausing. They also provide regulatory instruction in non-coding RNA strands. How did mere variation (noise) of duplication achieve such prescriptive sophistication? When have typographical errors ever improved a Ph.D. thesis?

20. How did codon-governed protein coding instructions get overlapped with Translational Pausing (TP) instructions by sextets of those same nucleotides? TP is critical to protein folding as the polyamino acid string emerges out the back door of the ribosome.[253,254]

21. How did inanimate nature know how to alternatively splice gene segments from different chromosomal locations into each needed gene? Inanimate nature cannot sense or respond to utilitarian needs. How could each alternative splice have been prescribed, managed and engineered by physicodynamic causation and/or heat agitation alone?

22. How and why did DNA mutation error-prevention/correction mechanisms arise out of mere chemistry?

23. How did isolated DNA methylations and histone acetylations, phosphorylations and methylations come to so tightly regulate gene switches into such exquisite and reliable development, and into such incredible holistic metabolic schemes?

24. Can inanimate nature formally organize anything? (Organization is quite different from unimaginative, redundant self-ordering such as crystallization, tornadoes, hurricanes, candle-flame shapes, sand-pile behavior.)

25. How did inanimate nature formally control events (as opposed to just constrain them)?

26. How did molecular evolution generate metabolic recipe and instructions using a representational symbol system? How did inanimate nature formally program using tokens in a material symbol system (MSS)?

27. How did inanimate nature send meaningful, functional messages governed by formal syntactical/linguistic rules?

28. How did physics and chemistry write and translate arbitrarily-generated code?

29. How did inanimate nature integrate conceptual circuits with insightful configurable switch-settings?

30. If choice-contingency is real at some level in animals, or even if just in *Homo sapiens* behavior, how did choice-contingency arise out of chance and necessity, mass and energy?

31. How did chance and necessity, mass and energy optimize algorithms? So-called evolutionary algorithms start with a pool of potential "solutions." But "solutions" are formal, not physical. A prebiotic environment would not have generated any "pool of potential solutions."

32. Suppose a self-replicative RNA analog sequence occurred spontaneously out of sequence space. How did this self-replicative strand simultaneously anticipate folding needs for metabolic utility? Any evolution toward *metabolic fitness* would tend to mutate the

Section 3: How did primordial prescription, and its processing, arise in a prebiotic environment?

sequencing away from *self-replicative fitness*. How could random mutations simultaneously contribute to both disparate functions?

33. How did so many biochemical pathways get integrated into one coherent, unified, symphonic-like concert of cooperative function (meta-metabolism)?

34. How did evolution, which has no goal, select components and manufacturing processes needed to realize the goal of metabolism?

35. What was the naturalistic mechanism by which highly-directional, multi-step biochemical pathways arose in cases where no selectable phenotypic benefit is realized until the final step in many of those biochemical pathways?

36. What selection *pressure* existed in a *prebiotic*, chemical-evolutionary environment to push physicality towards formal organization?

37. How was transitional connectivity (the "Continuity Principle" of life-origin research) achieved by nature between "protometabolism first" and "genetic takeover"?

38. How was homeostasis achieved—the maintenance of a relatively constant internal metabolic environment—despite changing external environments and thermodynamic decline?

39. How was homochirality achieved in protocells?

40. How was the difficulty overcome with which ribonucleotides are made?

41. How were ribonucleotides activated?

42. How was the problem of RNA instability overcome to provide the long periods of time needed for molecular evolution?

43. What was the driving force of *molecular* evolution?

44. How were exclusively 3'5' beta-D-ribonucleotide phospho-diester linkages universally established rather than 2'5' or 5'5' linkages?

45. How were deleterious cross-reactions avoided?

46. How was the hydrolysis of biopolymers prevented in an aqueous prebiotic environment?

47. If the prebiotic environment was non-aqueous, how did life get started in the absence of water?

48. Which of the four known forces of physics organized and prescribed life into existence? Was it gravity? Was it the strong or weak nuclear force? Was it the electromagnetic force? How could any combination of these natural forces or force fields program decision nodes to prescribe future utility?

49. There are an incredible number of all-or-none systems in even the most primitive cells. Are we certain that we have *fully* countered Behe's argument[336] regarding "irreducible complexity" in molecular machines and subcellular biosystems?

50. Have we demonstrated beyond all question that the "appearance or inference of design" in biosystems *can only be apparent* rather than real?

51. How empirically accountable, *in a prebiotic environment*, are the leading hypothetical models of abiogenesis (biochemically, thermodynamically, kinetically, physically)?

3.27.3 Queries needing answers for any naturalistic model of life origin to be plausible:

1. Does life require anything more than physics and chemistry?[30,473]

2. What is the basis for selection in a prebiotic molecular evolution environment?

Section 3: How did primordial prescription, and its processing, arise in a prebiotic environment?

3. Are molecular stability and longevity, along with self-replication adequate criteria for molecular evolution of all of the needed biochemical pathways/cycles and metabolism of life?

4. Can mass self- or mutual-replication of a couple of molecules generate and assemble at one place and time, in an organized fashion, all of the many hundreds of essential molecules needed for primordial life?

5. Wouldn't mass self- or mutual-replication of a few molecules have exhausted the remaining available resources from sequence space, so that none of the other hundreds of needed biopolymers would have been available?

6. How did chance and necessity, mass and energy in a prebiotic environment generate the first *control* mechanisms?

7. Could life exist without information?[4,6,7,23,30,34-38,50,71,81,85,172,346,359,409,474-487]

8. What kind of information does life require?[4,6,7,23,30,36-38,50,81,85,172,346,359,477-485,487,488]

9. Is the mathematical definition offered for "functional information"[4] adequate to *program/instruct/prescribe* genomic and epigenomic organization and control?

10. Is information physical?[6,7,21,22,29-31,42,61,68,70,71,84-86,489,490]

11. How did chance and necessity, mass and energy program metabolic success?

12. How did the redundant, monotonous, unimaginative, dissipative structures of chaos theory organize such sophisticated, integrative, biochemical utility needed for protolife?

13. Can mere physicodynamic constraints optimize algorithmic controls?

14. Could the mere circular constraints of hypercycles have generated formal organization, control and such finely-tuned epigenetic regulation?

15. Why have there been absolutely no observations or prediction fulfillments of hypercycles ascending spontaneously into Sustained Functional systems (SFS) since Eigen's first hypercycle publication?

16. Did the first *cell* require Prescriptive Information (PI)?

17. Did the first *protocell* require Prescriptive Information (PI)?[21,22,71]

18. How was each step of *new* instructions/programming prescribed leading up to abiogenesis?

19. Duplication of what? How did the initial Prescriptive Information (PI) that is being duplicated come into existence? When we talk about duplication plus variation, aren't we just presupposing, rather than explaining, the initial PI that is being duplicated?

20. Just how much sophisticated new function can we expect *random* variation of duplication to produce?

21. Just how much sophisticated new function can we expect *non-random* mutations of duplicated PI to produce? A significant number of mutations are now known to be non-random.[198,491-496] But how many creative innovations could mere constraints and fixed law have programmed?

22. If noisy variation of duplicated Prescriptive Information (PI) was all that was needed to generate so much new sophisticated PI, why do we have to go to such great lengths to prevent noisy variation in Shannon channels in our everyday communications?

23. If duplication plus mere variation is the source of all improvement in genetic and genomic programming, wouldn't noise variation in Shannon channels improve our daily communications?

Section 3: How did primordial prescription, and its processing, arise in a prebiotic environment?

24. Programming (e.g., of DNA) requires freedom of selection at bona fide decision nodes, logic gates and configurable switch settings. How did a prebiotic environment establish this freedom from constraints and law?

25. How were functional nucleotide selections, along with nucleotide syntax, made in advance of the realization of any utility?

26. Natural selection (NS) favors only the fittest *already-programmed, already-computed, already-living* phenotypes (The Genetic Selection [GS] Principle).[71] How did NS work at the nucleotide selection level during prebiotic polymerization?

27. How did a passive, protocellular, lipid, pseudo-membrane simultaneously acquire discriminatory active transport capable of 1) managing osmotic gradients, 2) ingesting needed nutrients, 3) excluding potential environmental toxins, and 4) excreting so many different kinds of metabolic wastes?

28. Why has empirical evidence of spontaneous *formal* self-organization been *so* absent in nature over the last 140 years, despite every biologist on earth looking for that empirical evidence?

29. The programming of formal function is goal-oriented. But physicodynamics and evolution have no goals. So how could potential function have been pursued by protocells in a prebiotic environment?

Of course, we still have many unanswered questions about homochirality, RNA instability, the extreme difficulty of making cytosine, especially in a prebiotic environment, etc. All of these problems are yet to be resolved.

In the end, we of course need to apply all of this logic to the origin of subcellular cybernetics in a prebiotic environment. But before muddying the water, we need first to make sure everyone is on the same page with regard to the necessity of Selection *FOR (in pursuit of)* in order to prescribe or program. Cyberneticists of any kind need to be cornered with the simple fact that chance and necessity cannot program anything. Then, naturalistic scientists, biologists especially, also need to be cornered with the same reality.

Critiquers of this book are invited to challenge the validity of the following previously published fundamental dichotomies:

A. Nonphysical formalisms (e.g., mathematics, language, programming, coding, translation, linear digital prescription) vs. physical chance and necessity, mass/energy interactions.[19,31,300-303]

B. Formal controls vs. mere physicodynamic constraints.[61]

C. Spontaneous self-ordering in nature vs. formal organization.[7]

D. Dissipative Structures of Prigogine[288] vs. Sustained Functional Systems[30]

E. Selection *FROM AMONG existing* fitness vs. Selection *FOR potential* fitness (needed to program both genomics and epigenomics).[20,22,31,42,83]

F. The mere physics definition of "work"[300-303] vs. the intuitive utilitarian definition (the useful, pragmatic "work" performed on countless levels by even the simplest living cell).[6,19,29-31]

G. Mere negative Shannon uncertainty vs. positive Prescriptive Information (PI).[5,71]

H. Physicodynamic causation vs. Choice-Contingent Causation and Control (CCCC) (also published under the name Choice Determinism [CD]).[19,41,42]

3.28 Primordial life would have required Prescription and its Processing.

Even Protocells would have needed instructions, organization, controls and regulation. First, PI was needed. That PI also had to include the "know-how" for processing those instructions. "Know-how" for processing, however, does not measure up to the actual processing of PI. All of the computing components had to be there, in place, ready to process. Those components themselves had to be programmed to process PI instructions. Actual physical ribosomes, for example, would have been needed.

All known life is universally prescribed with genomic and epigenomic controls, not just constraints. Life, development especially, is mediated through programmable configurable switch-settings. Any naturalistic life-origin model must address the fact of this prescription, and how it and its processing could have emerged.

3.28.1 Prescription and its Processing (P & P) are life's most essential ingredients.

Subcellular programming employs Ontological Prescriptive Information (PI_o) rather than Epistemological Descriptive Information (DI_e).[5,23,29] Biochemical pathways are steered toward metabolic success independent of human knowledge. Cycles contribute to the holistic goals of free energy generation, transduction, storage, availability when needed, and actual utilization. Staying alive and reproducing are the ultimate teleological quests. All other notions of teleology depend upon these subcellular ontological teleology's. Trying to naturalize them merely be changing their names to "teleonomy's" will never change their most fundamental properties—choice-contingent steering of events toward needed utility—staying alive and propagating. Ontological Prescriptive Information (PI_o), and its processing, governs all of these capabilities and processes.

PI_o steers and regulates the metabolism of every known organism. Metabolism is invariably controlled by genomic and epigenomic PI_o

programming. PI_o's existence would have had to precede life's existence, not be its product. It could not have been born simultaneously in conjunction with abiogenesis, either. Prescription precedes life's essence. Instructions, programming and algorithmic processing were all needed to *cause* the effect of "life."

Many thermodynamic, chemical and biophysical questions remain to be answered in life-origin science.[497,498] Elucidating the origin of PI in a prebiotic environment, however, remains by far the most fundamental and daunting challenge of naturalistic abiogenesis research.

If prescription was required for life to become organized and programmed, how did prescription arise in a prebiotic environment? What are the necessary and sufficient conditions for prescription to exist? If prescriptive information spontaneously arose first, how were these instructions conceived and recorded? How did such instructions get translated into an actionable format? How was this prescriptive information processed into a useful product? Wouldn't machines, or nanocomputers of some sort at the subcellular level, have been required to process such computational programming? What does inanimate nature know about "instructing," "processing," "usefulness," or "computation"?

Given the Second Law, no "metabolism first" model could possibly have sustained itself for long without acquisition of a heritable instruction set and information processing capability. The metabolism wheel could not have been re-invented spontaneously many times. The probability of a first-time "self-organization" of metabolism is statistically prohibitive as it is. In addition, any temporary and local movement away from equilibrium could not have been achieved or maintained without steering. Formal controls would have been needed above and beyond mere physicodynamic constraints.[61] Maxwell's demon achieved useful energy potential for the simplest conceivable heat engine *only through choosing* when to open and close the trap door.[5] Any attempt to deny the role of Controls, as opposed to mere constraints, results in an immediate return to movement towards equilibrium between the two compartments. There is only one answer as to what a childish cartoon of a devil is doing in virtually every physics text: There was no escaping choice-contingency operating the trap door if an energy potential was to be created with inert gas molecules. Naturalistic physics and chemistry could never explain that phenomenon.

The key to life is Ontological Prescriptive Information (PI_o) and its algorithmic processing, both of which depend upon formal, nonphysical CCCC.[19,29,31,42,499] Programming choices are nonphysical and formal. Biological PI_o and its formal processing are only two of many pieces of evidence in support of the most fundamental principle or axiom of science, *The Formalism > Physicality (F > P) Principle*.[19,29] (See Section 3.16)

Life cannot exist without organizational concept and impetus. This requires foresight and purposeful programming choices at bona fide decision nodes.[2,5,20,39,70,71,79,86] Life also requires tight controls at every turn.[500-503] Travel down certain pathways must be steered and integrated with hundreds of other pathways.

The key to understanding life is not negative Shannon uncertainty, combinatorial "complexity," or thermodynamics. The key to understanding life is the positive *"prescription and optimized continuing regulation of bio-function."*[6,19,21,22,29-32,39,42,62,82,83]

Error correction, another form of PI, is also involved.[310] Prescription of function provides a form of utilitarian certainty, not uncertainty. But since biological information is ontological, we cannot really talk in terms of certainty any more than we can of Shannon uncertainty. Even if we could, merely subtracting one uncertainty from another uncertainty does not generate or explain such exquisitely imaginative prescription of bio-function as we observe in the simplest cell. Nor would discussions of human "reduced uncertainty" answer the question of *what* reduced the cell's "uncertainty," (as if the cell had a brain and mind of its own).

The varieties of functional controls of both small and large ncRNAs have quickly become mind-boggling.[504] "Selection *FOR* potential function" at bona fide decision nodes (literal configurable switches) is required to generate PI.[2,6,7,20,39,42,82] Decision nodes exist throughout the cell in various forms, including Selection *FROM AMONG* symbols from alphabets of symbols, configurable switch-settings, the implementation of ingenious cooperative strategies, and the integration of component parts into holistic machines.

The most prominent decision-node selections occur at the programming level of nucleotide sequencing.[2,6,7,20,26,27,39,42,82,315] This is the equivalent of writing linguistic syntax or programming machine code onto a theoretical Turing tape. But the mere "double helix" of DNA structure does not provide the prescriptive data feed that organizes life. Particular nucleotide sequencing carries considerable sophisticated instructions and controls.[2,6,7,20,39,42,61,82] DNA, in a very real sense, even provides hardware capability.[23,24,26,27,315] Physicodynamically-indeterminate nucleotide syntax also determines ribozyme structure and function. Nucleotide syntax also prescribes genetic and ncRNA regulatory prescription.[44,45,59,505] DNA is read in both directions.[251,506-510]

3.28.2 How does Choice Determinism contrast the spontaneous generation of life?

In the introduction page of this book, we talked about Choice Determinism (CD) being distinguished from law-like physicodynamic and Physicochemical Determinism (PD) in two major ways. First, CD is never automatic or spontaneous in inanimate nature. CD does not and cannot arise from mere chance and/or necessity (natural law). Second, CD can only arise from purposeful choices that steer or direct behavioral outcomes toward some desired function.

CD is not always pragmatically wise. But CD is normally exercised under the desire and belief that purposeful choices will yield beneficial results, as valued by some agent. Only "agents" are known to value anything. Only agents are known to pursue attainment of such value. We then defined and explained agency as viewed by philosophers of science. In all naturalistic science views of the history of presumed objective reality, no agency existed in prebiotic nature.

Yet agency is the only known participant in valuation and pursuit of functionality and "useful work." Thus, in the absence of any agency in a prebiotic environment, nothing existed to "prefer" or pursue function over non-function.[370,511] Even today, the environments of inanimate nature neither value nor pursue formal utility. An inanimate environment would be utterly blind to the attribute of "usefulness."

Any theory of natural selection depends squarely upon the attribute of "usefulness," "preferential function," or "superior fitness." If an inanimate environment cannot even detect, and does not care about, "fitness," it cannot possibly prefer or select for greater "fitness" at the programming level. A "fitness function" (e.g., in genetic algorithms) cannot possibly exist independent of valuing and preferring agents. Yet the illusion is constantly created at the hands of metaphysical naturalists that artificial genetic algorithms are modeled after inanimate nature's own "natural genetic algorithms." No such thing exists in inanimate nature. The entire concept of natural genetic algorithms is a ruse—a shell game. Naturalistic chemical evolution[512] is totally without empirical or rational support. Zero empirical or rational evidence exists for any genetic algorithm existing spontaneously in nature apart from agency.

This point is never so clear as when we begin our modeling of life-origin in a prebiotic environment, where nothing exists but inanimate nature. It is from this purely inanimate state that we must create a plausible theoretical explanation for the emergence of bona fide organization (not mere self-ordering of Prigogine's dissipative structures). We must explain from a prebiotic environment the emergence of genomic controls, even if those controls are instantiated into some physical analog of RNA rather than DNA.[513] But controls, as opposed to mere physicodynamic constraints, only come into existence at the hands of agency, not inanimate prebiotic nature. No goals exist in prebiotic nature. No function is valued over non-function. An inanimate prebiotic environment just IS. It has no preferences or desires. It has no intentionality. Absolutely no impetus exists for the spontaneous generation of life. No reason exists for the organization of the simplest biochemical pathway, let alone integration into a network of cooperating pathways and cycles, that would all contribute to the common goal of "metabolism." Controls are always the subset of Choice Determinism (CD), not Physicodynamic Determinism (PD).

The most elementary cell relies heavily upon extraordinary controls and regulation. No basis exists for trying to argue that primordial cells would not have needed any of these controls. We can plausibly entertain that controls would have been far simpler than those in current life, but we cannot maintain that controls were unnecessary for protocells and primordial cells to organize and "come to life." Too many different biochemical pathways must be directed toward cooperative integration. Too many directional controls and too much precise enzymatic regulation are required for even the simplest metabolic cycles to organize and be maintained, let alone the simplest holistic life (e.g., *Mycoplasma genitalium,* or maybe *Carsonella ruddii*[514]).

3.28.3 Biological PI_0 is nonphysical, the same as any other PI.

In molecular biology, "The 'meaning' (significance) of prescriptive information is the function that information instructs or produces at its metabolic destination."[6] Szostak has used the term "functional information."[37] Prescriptive Information (PI) includes instruction and algorithmic/computational programming. Descriptive Information (DI) contains only epistemological description. Genes provide ontological instructions, algorithmic prescription and programming of computational function.

The oft used term "complexity" in life-origin literature is grossly inadequate to define the nature of genetic control.[70,79,81,85,86,275] As

Hoffmeyer and Emmeche point out,[50, pg. 39] "Biological information is not a substance." Later they repeat, "But biological information is not identical to genes or to DNA (any more than the words on this page are identical to the printers ink visible to the eye of the reader). Information, whether biological or cultural, is not a part of the world of substance."[50, pg. 40] As stated earlier, the formal, nonphysical, prescriptive selections instantiated into configurable switch settings (nucleotide selections in this case) must never be confused with the physicality of those configurable switches themselves.

The maximum length of oligoribonucleotides in aqueous solution is only 8-10 mers.[515] The genetic programming of longer strands is certainly not "blind." Stochastic ensembles of single-stranded small RNAs or of polyamino acids do not fold into functional shapes. Yet both nucleotide and dipeptide overall frequencies seem close to random in living organisms.[516,517]

Biomessages are unique in nature in that they are formally and functionally sequenced.[518] They are not randomly sequenced, and they are not ordered by physical laws. They are sequenced so as to encrypt programmed instructions for the undeniable, far-removed goal of achieving overall homeostatic metabolism. The realization of this goal requires transcriptional editing, decryption (translation), chaperone assistance in folding, and sometimes even post-translational editing.[519] These processes are fundamentally formal, as formal as the mathematical laws of physics. The genome and its editing processes not only prescribe, but directly and indirectly compute the end product.

In a Peptide or Protein World model of life origin, efficacious Selection *FROM AMONG* each amino acid must be explained at the level of covalent peptide bond formation. Polyamino acid primary structure (sequence) is formed prior to folding. Primary structure is the main determinant of how the strand will fold. Thus, functional shapes must be prescribed by linear digital semiosis. The covalent bonds of these highly informational strings are "written in stone" prior to when weak hydrogen-bond folding secondarily occurs. Instructive sequencing must be completed before tertiary shape and function ever occur. The GS Principle, or Genetic Selection Principle, holds true. This principle[6,71] states that selection must operate at the genetic level, not at the phenotypic level, to explain the origin of genetic prescription of structural and regulatory biological function. This is the level of configurable switch-settings (nucleotide selection). Selection must first occur at each decision node in the syntactical string. Initial programming function cannot be achieved by chance added to after-the-fact Selection *FROM AMONG* the already-existing fittest programs ("phenotypes"). Evolution is nothing more than differential survival and reproduction of already-programmed, already-living, fittest phenotypes. *The computational programming proficiency* that produced each and every phenotype must first be explained. Programming takes place at the genetic and epigenetic level. Even epigenetic prescription, development, and regulation ultimately trace back to the genetic programming of those sRNAs and regulatory proteins. Thus far, no natural-process explanation has been published to explain Selection *FOR (in pursuit of)* at the decision-node, configurable-switch, nucleotide-selection level. The firm prediction is made by this author that none will be, because, logically, none *can* be. Chance and necessity cannot Select *FOR (in pursuit of)*.

Even the translated polyamino acid language is physically nonfunctional while forming. Not until after it dynamically folds according to the instructions contained within its linear digital programming (its primary structure) does it become functional. Translational pausing at the back end of the ribosome also plays a role, as determined by hexanucleotide coding superimposed on the triplet codon coding.[253,254] Only later does this syntax of covalently (rigidly) bound monomeric sequencing determine minimum-Gibbs-free-energy folding. Highly prescribed chaperone protein molecular machines are also often needed to fold proteins correctly. Even then, not even three-dimensional shape, or tertiary structure, is selectable by the environment. A far more holistic context of differential organismic survival and reproduction are required for natural selection to kick in.

In molecular biology, recipe code is translated from nucleotide sequence language into a completely different conceptual amino acid language. A correspondence exists of representational meaning between arbitrary alphanumeric symbols in different symbol systems. Each triplet codon is a Hamming "block code" for a single letter (amino acid) of a long protein word.[439] A prescriptive codon prescribes a certain amino acid letter at the receiver upon decoding. It is often argued by philosophic naturalists that the symbol system and code translation of molecular biology are only heuristic. But the prescriptive role is quite real. The correspondence between the codon-block-code sequencing and amino-acid sequencing is clearly not physicalistic. Nucleotide sequencing is physicodynamically arbitrary and resortable. Coding is formal, not physicodynamic. No binding or physicochemical reaction occurs between nucleotide symbols and the amino acid symbols they represent. Anticodon and amino acid are on opposite ends of each tRNA. Aminoacyl t-RNA synthetases are also independent enzyme molecules that have no direct binding affinity to codons. Neither fixed laws nor chance-contingency can explain the integration of twenty different amino acids, aminoacyl t-RNA synthetases, the specific amino-acid ends of each tRNA molecule, the specific anticodon opposite ends of each tRNA, and the Hamming "block code" of each triplet codon. The number of permutations of polyamino acid strings is staggering. The spontaneous integration of all these individual entities into a formal association capable of promoting even a protometabolism is not only statistically prohibitive, but violates the Universal Plausibility Principle (UPP).[80] Just as physics is governed by formal computation, biology is also governed by formal computation. Both require traversing the Cybernetic Cut to describe, understand, and create predictive models.

Recent peer-reviewed papers are beginning to refer to "molecules of choice."[520] It is fast becoming obvious that all sorts of additional kinds of information transfers, besides linear digital genetic instruction, control every aspect of subcellular metabolism.[521] Choice Determinism (CD) alone explains the phenomenon of ontological PI and cybernetic processing.

3.28.4 Life-origin is about the emergence of P & P.

It is helpful to keep emphasizing "Prescription" more than "Information" in the study of abiogenesis.[2,6,7,20-22,31,39,42,62,70,82] Talking about "information," even Functional Information (FI), tends to steer the discussion back into the quagmire of human epistemology. If there is any hope of coming to grips with the reality of a non-subjective, ontological PI_o, it lies in studying the stand-alone, objective, prescriptive control and processing phenomena of protocells and primordial cells.

Biological PI_o organized not only the first prokaryotes, but the first protocells. No protocell theory can get off the ground without at least elementary goal-oriented controlling systems that pursue circuit integration, productive biochemical pathways, cycle organization, positive and negative feedback mechanisms, purposeful energy harnessing, storing, transducing and utilizing cybernetics.

No basis exists for undying faith in spontaneous self-organization, or for relentless climbs up to mountain peaks of rugged fitness landscapes. Any envisioned naturalistic hypercycle model is unfounded. No naturally-occurring hypercycles have occurred that produce bona fide organization or non-trivial function. Tightly regulated genetic control and meta-metabolism simply do not occur. The notion springs only from blind faith, not scientific empiricism. So-called "self-organization" has not only never been observed, it is logically self-contradictory.[5,6,29,30,62,85] Nothing can organize itself into existence. An effect cannot be its own cause. While physicodynamics can spontaneously self-order to form dissipative structures, only formalisms can organize anything. Experiments that appear to generate FI spontaneously, and to organize physicality, are invariably artificially steered toward optimization by experimental design. The experimenter chooses and pursues the outcome in direct violation of evolution theory. "Directed evolution" is also a self-contradiction.

Even a prokaryote's choice of approaching a higher glucose concentration, or avoiding a higher concentration of environmental toxin, requires an incredible amount of subcellular prescription of metabolic organization and function (Ontological Prescriptive Information [PI_o]). Whether we view the prokaryote as a completely preprogrammed robot, or as a robot with considerable preprogrammed "end-user degrees of freedom," this cell is still ruled by mind-boggling PI_o. This PI_o has never even been addressed, let alone explained, by naturalistic life-origin science. This blatant effect must have a cause according to naturalistic cause-and-effect science. What is that cause? How can mere chance and necessity—mere mass, energy, force fields and heat agitation—generate wise programming decisions instantiated into physicodynamically indeterminate nucleotide selection and sequencing? How did all the cytosine methylation configurable switches, for example, get set?

These formalisms can be instantiated into physicality using tokens in material symbol systems[43-45,522] and physical configurable switch settings.[6,21,22,29-31,42,61] But, the functionality of these physical instantiations of nonphysical formalisms is made possible only by formal instructions originating from the far side of The Cybernetic Cut[42,70,299] across the one-way Configurable Switch (CS) Bridge.[41,42,299] Chance and necessity cannot generate Choice Contingent Causation and Control (CCCC) or Choice Determinism (CD). Yet, CD is necessary for any form of programing and computational success.[19,29,31,42] All known life is highly programmed and computational.[29]

Physicodynamic Determinism (PD) and Choice Determinism (CD) are in two different fundamental categories. Anyone working within naturalistic presuppositions is limited to only two categories: chance and necessity. There is a third undeniable category: Selection (even to believe in macro-evolution!). But when we subdivide Selection into its two subsets, and understand the difference, we see immediately that evolution is limited to Selection *FROM AMONG*. Evolution cannot Select *FOR* (in pursuit of) potential function (See The Universal Selection Dichotomy [USD], Section 3.3).

All of the pieces of the objective reality puzzle cannot fit within the perimeter of the "chance and necessity only" worldview.

Since all known life is cybernetic (programmed), evolution is worthless in explaining any increase in PI_o at ANY stage of macro-evolutionary theory.

No empirical or rational basis exists for anyone believing in a spontaneous increase in PI_o from natural causes.

This is not just true with abiogenesis. It is also true of macro-evolutionary theory. We can observe variation of existing PI with a slight variation of existing prescription. But we will not see spontaneous increases of prescription of sophisticated function arising from the near side of the Cybernetic Cut. Chance and necessity, mass and energy, will not act on existing genomes and epigenomes to increase their programming proficiency!

This book began with, "It's not about information. It's about PRESCRIPTION, and the processing of that prescription." Think of information as a Turing tape. What good is a Turing tape without processing? Processing usually requires sophisticated machinery or computer-like devices. Those devices require a great deal of Prescriptive Information (PI) and processing themselves.

It's about CHOICE DETERMINISM (CD) in pursuit of non-trivial function. Disallow Purposeful Choice, and you disallow Prescription and its Processing! What matters is the actual *causation* of sophisticated formal function, and the integration of many bio-functions into holistic metabolism. Mere information doesn't DO anything.

Section 3: How did primordial prescription, and its processing, arise in a prebiotic environment?

The problem of macroevolution is no different from the problem of abiogenesis in the final analysis: Chance and necessity, mass and energy, cannot program new computations, or generate any new abstract, conceptual, formal instructions to prescribe new non-trivial integrative function.

The only suggestion of empirical evidence for "accumulation of bio-information" comes straight from a confused definition and misunderstanding of what PI really is. There are so many polluted definitions and ideas of "information" floating around that we cannot trust any discussion of "information." PI_o is not spontaneously accumulating, even in existing living systems. Existing genomes are under no circumstances generating new instructions for more sophisticated genera to come into existence. And they are certainly not producing more sophisticated genera. No new PI is generated from duplication plus noise variation of existing genomes. The only route to more sophisticated programming is through Choice Determinism (CD), not Physicodynamic Determinism (PD) or chance-contingency.

CD is required to Select *FOR* (in pursuit of) any potential function that does not yet exist for the environment to prefer. Probabilism cannot address or measure Decision Theory and CCCC.

DNA and highly edited mRNA are Turing tapes "written in stone," with strong covalent chemical bonds, long before those sequences can be translated into polyamino acid strings. Even after they are finally translated into a completely different language of polyamino acid strings, those strings only fold up into the badly needed three-dimensional protein molecular machines on the basis of minimum-free-energy thermodynamics. The latter is purely physical. But the latter is *determined* by the sequencing of amino acids that was prescribed by the codon sequence in mRNA. The codon sequencing is formal, not physical. All of the chemical bonds between codons are the same. The sequencing is not determined by the laws of physics. If they were, DNA would consist of all the same letter (nitrogen base). We would have a polyadenosine homopolymer of all the same letter—

(A for adenosine)—by law! The message would be meaningless, the same as if you had a computer program of all 0's, or a computer program of all 1's.

What makes programming, and also life, possible is *freedom from law!* Contingency must exist. But there are two kinds of contingency: chance-contingency and choice-contingency. No programmer ever produced a functional, computational, halting program by flipping a coin at every decision node. Many programs will completely crash because of one bad choice at one decision node in the entire program! Clearly, neither chance nor necessity (law) can explain the programming of life. Yet chance and necessity, mass and energy, are all philosophic naturalism has to work with to explain anything.

Once in a while a random sequence can *resemble* a recognizable word (to turn the tables on Dawkins, "have the appearance of being a word)". But that does not change the fact that the sequence is random and useless until an *agent's mind reads into that random sequence his own lexicon and arbitrary rules of language*. The agent then thinks in his own mind that a meaningful word has accidently occurred, but it hasn't. It's still a random sequence. It's like lying on your back in the yard and watching clouds go by. Sooner or later your mind will "recognize" (read something familiar into) a random cloud formation. But the cloud formation itself is still random and meaningless.

Not so with genomic prescription. Genomes *objectively*, *ontologically* prescribe exquisite function. No human epistemology or imagination is involved. Genomic prescription of sophisticated computational function not only predates humans, it produced humans. Specific switch-settings also determine how RNA strands fold back onto themselves, forming helices, bulges, loops, junctions, coaxial stacking, etc.[515, pg. 682-683] Not even the hypothesized pre-RNA World and RNA World[255] escape the formal linear digital algorithmic governance of computational function. The generic chemical properties alone of nucleic acid and protein are insufficient to generate life.

All empirical and rational evidence in the history of human observation testifies to the fact that prescription of sophisticated function has never arisen independent of CD. Computational linear digital algorithms distinguish life from non-life more than any other singe characteristic.[523][524] Says Yockey,

> The existence of a genome and the genetic code divides living organisms from non-living matter. In living matter chemical reactions are directed by sequences of nucleotides in mRNA. . . . There is nothing in the physico-chemical world that remotely resembles reactions being determined by a sequence and codes between sequences.[525, pg. 54]

Küppers[526, pg 166] makes the same point as Jacques Monod,[40] Ernst Mayr,[271,272] and Hubert Yockey.[209,439] Physics and chemistry do not explain life. Niels Bohr argued that "Life is consistent with, but undecidable from physics and chemistry."[305] What exactly is the missing ingredient that renders life unique from inanimate physics and chemistry? The answer lies in the fact that life, unlike inanimacy, originates from the far side of the Cybernetic Cut.

3.28.5 The worship of "possibility"

It is incumbent upon any physicalist to explain how mere "possibility" magically got converted into programmed, formal computational success. Philosophic naturalists tend to worship "possibility" because they have no choice. "Possibility" is *all* they have to work with. Unfortunately, the statistically prohibitive "possibilities" that philosophic naturalism desperately clings to usually have Universal Plausibility Metrics ξ (ksi) of < 1.0. Any model of causation with a $\xi < 1.0$ is considered definitively falsified by scientific methodology. The proposed model should be rejected by peer review in all fields of science in accord with the Universal Plausibility Principle.[80,330]

An infinite number of universes and quantum quackery are next introduced in a feeble attempt to try to explain the undeniable, repeatedly observable, FACT of cybernetics in subcellular reality.

Blind belief in an infinite number of universes is pure metaphysics. It is completely outside the bounds of empirical science. Until actually observed, it is nothing more than science fiction.

Multiverse models imagine that our universe is only one of perhaps countless parallel universes.[527-529] Appeals to the Multiverse worldview are becoming more popular in life-origin research as the statistical prohibitiveness of spontaneous generation becomes more incontrovertible in a finite universe.[530-532] The term "notion," however, is more appropriate to refer to multiverse speculation than "theory." The idea of multiple parallel universes cannot legitimately qualify as a testable scientific hypothesis, let alone a mature theory. Entertaining multiverse "thought experiments" almost immediately takes us beyond the domain of responsible science into the realm of pure metaphysical belief and conjecture. The dogma is literally "beyond physics and astronomy," the very meaning of the word "metaphysical."

The notion of multiverse has no observational support, let alone repeated observations. Empirical justification is completely lacking. It has no testability: no falsification potential exists. It provides no prediction fulfillments. The non-parsimonious construct of multiverse grossly violates the principle of Ockham's (Occam's) Razor.[533] No logical inference seems apparent to support the strained belief other than a perceived need to rationalize what we know is statistically prohibitive in the only universe that we *do* experience. Multiverse fantasies tend to constitute a back-door fire escape for when our models hit insurmountable roadblocks in the observable cosmos. When none of the facts fit our favorite model, we conveniently create imaginary extra universes that are more accommodating. This is not science. Science is interested in falsification within the only universe that science can address. Science cannot operate within mysticism, blind belief, or superstition. A multiverse may be fine for theoretical metaphysical models. But no justification exists for inclusion of this "dream world" in the sciences of physics and astronomy.[80, pg 7]

Quantum events are statistical. This means there is no basis in quantum theory for purposeful "Selection *FOR (in pursuit of)*." Even if statistical theory could somehow account for purposeful prescriptive choices, we would still be dealing with a *formal* subcellular computational process that denies naturalistic philosophic presuppositions ("Physicality is sufficient."). Statistics don't *cause* any physical effects. Statistics are purely descriptive, not prescriptive. Quantum theory explains absolutely nothing about the derivation or ongoing cybernetics of any cell.

3.28.6 Still another chicken-and-egg paradox to life origin

P & P is life's most essential ingredient. This P & P is ontological. It began in a prebiotic environment. It had, and still has, objective existence independent of human knowledge or understanding. That is to say, the P & P that makes life possible is not epistemological, although we are learning more *about* it by the month in scientific literature.

Life originated only out of formal organization, programming choices, steering, controlling instructions, extraordinary regulation, and ongoing processing. Yet, no source of this P & P is known other than life itself. Thus, we have still another chicken-and-egg paradox associated with life-origin studies.[534] This ontological P & P (P & P_o) paradox is by far the most perplexing of all life-origin naturalistic paradoxes. Only agency can make programming decisions and engineer processing devices. Only agency can organize parts and events into useful entities and procedures. Yet agency exists only in association with, or as a product of, already-existing life.

Which came first, the chicken or the egg? *It doesn't matter*. Both the chicken and the egg are highly prescribed. Both are replete with extraordinary formal organization (not just physicodynamic self-ordering). This organization includes: purposefully set configurable switches; integrated circuits; computationally successful schemes; developmental plans; end-user programmable preferences; contingency modules that prescribe appropriate responses to almost any environmental stress or opportunity. All of this genomic and epigenomic programming predates and "makes happen" phenotypic cells. This programming and processing alone compute and manufacture organisms with superior fitness. Environmental selection is only secondary—after the fact of this programming and processing at the decision-node level of genomic and epigenomic prescription.

What does naturalistic science do with so many different life-origin paradoxes? Most have been well known and fully appreciated for over fifty years. Usually, naturalistic science quickly sweeps them under the rug.[535,536] Of utmost importance to philosophic naturalism, and to any other presuppositional belief system, is to protect and maintain its starting axioms. All thinking must flow deductively from those axioms. But, when massive observational and logically consistent evidence mounts that cannot possibly flow from those axioms, it is time to consider the possibility that those beginning axioms of that metaphysical commitment might have been mistaken—inapplicable to the real world in which we all have to live. We have to consider the very real possibility that our current philosophy simply does not *correspond* to objective reality.[537] At this point, sometimes an extremely rare, major, Kuhnian paradigm shift[538] is necessary in science. We have come to such a point, particularly in the science of biology. But a stubborn refusal to acknowledge the facts prevails.

It is interesting to note that every last chicken-and-egg paradox of life-origin science immediately resolves the minute materialistic metaphysical presuppositions are questioned. Science prides itself in skepticism. Why do we choose to suspend all skepticism in order to maintain the purely philosophic dogma of physicalistic naturalism?

Section 3: How did primordial prescription, and its processing, arise in a prebiotic environment?

Elucidating the origin of Prescription of Function (PoF) first entails explaining the phenomenon of true organization. Organization is not just the self-ordering described by chaos theory. Bona fide "organization" requires purposeful choices in pursuit of utility. The dissipative structures of chaos theory (e.g., hurricanes, tornadoes, the shape of candle flames, sand pile behavior) never involve purposeful choices. Scientific literature is replete with utter confusion that stems directly from failure to realize and understand the clear dichotomy between "organization" and "self-ordering."[31,62,79] The result is endless propagation of nonsense notions such as "self-organization." Not even humans, with all of their intelligence, could logically "self-organize" themselves into existence. An effect cannot be its own cause.

Resistance to paradigm change can be both a blessing and a curse to science.[538] The number and severity of chicken-and-egg paradoxes associated with naturalistic life-origin questions have been so painfully apparent for so long that no excuse remains for obstinate refusal to reconsider purely metaphysical naturalistic presuppositional commitments. Dogmatic philosophic belief in physicalism is long since utterly bankrupt, especially in view of the undeniable organizational and controlling role of nonphysical, formal PI_o in all living organisms. The formal nature of the needed processing of PI_o only makes the physicalistic axiom all the more ridiculous. It is well past time to question the incorporation of naturalism into the very definition of science. Not even scientific method itself is "naturalistic." There is nothing physical about the mathematics upon which physics depends.

3.28.7 The multi-dimensional/multi-layered nature of biological prescription

Prescriptive choices are not always linear, sequential, one-dimensional, or single-layered. A fundamental prescriptive choice can radiate into multiple dimensions and layers of CD and PI. Thus, the choice of a nucleotide at a certain locus in a polynucleotide linear-digital string provides the needed CD and PI_o. Complimentary base-paired strands can simultaneously contain epigenetic switches that control genes and histone codes, for example. Not just mRNA, but short and long ncRNAs are managed.

The choice of a nucleotide itself is not the fundamental biological unit of choice. Not only is it a quaternary rather than a binary choice, but the choice can be further subdivided according to multiple dimensions or layers of function. The nucleotide's methylation or acetylation, for example, can quickly double the functional choice opportunities at that locus in the DNA string. Both methylation and acetylation can serve to switch on and off many sophisticated genomic and developmental functions. The question for programming the system is not just which nucleotide to "choose," but whether or not a cytosine should be methylated. Then, there are histone programming options in additional layers and dimensions of PI. But, the three-dimensional tertiary structure of the histone proteins can be determined by linear digital programming of the protein's primary structure.

What was thought to be merely redundant codons, all prescribing the same amino acid, are not redundant at all in the larger picture. Hexanucleotide prescription of Translational pausing (TP) and protein folding overlaps, and is superimposed onto, codon prescription of each amino acid. Not just any codon will do in prescribing the same amino acid, as previously thought.[253,254] Happenstance could not possibly have prescribed such overlapping layers and dimensions of PI_o and its processing.

3.28.8 The first P & P's were ontological, not epistemological.

Biologists desire to know what specific polymeric sequences will optimize metabolic schemes. We spend billions of dollars to ascertain "reference sequences" (the family of specific sequences capable of contributing to metabolic function). The acquired knowledge of these specific sequences of "instructions" is the essence of what we call meaningful (semantic and pragmatic) information. The difference between this kind of information, and that found within cells, is that the Prescriptive Information (PI) found in cells has nothing to do with human knowledge.

Section 3: How did primordial prescription, and its processing, arise in a prebiotic environment?

Polymer strings provide linear digital prescription of eventual molecular folding function that has nothing to do with human knowledge. Subcellular Turing tape-like instruction, and its Turing machine-like algorithmic processing, predated Turing, Church, Weiner, and every other cyberneticist. Human minds can come to learn a great deal *about* subcellular PI. But the PI itself is in no way dependent upon human knowledge to "do its thing." It was obviously organizing, controlling and regulating life long before *Homo sapiens* came into existence to discover anything about it. Whatever cellular P & P is, it cannot be reduced to mere human epistemology. And it certainly cannot be reduced to any modified statistical measure. It cannot be reduced to chance and necessity, mass and energy, either.

The minute we specify the family of which particular strings "work" in any biological scheme, we have already provided detailed epistemological Functional Information (FI) *about* that Sustained Functional System (SFS). We have provided extrinsic information from outside the system *about* what is needed to prescribe and process it.

But, *aboutness* did not exist in a prebiotic environment as envisioned by naturalistic metaphysics. No central nervous systems existed on earth at the time to "know" anything. No organismal brains, minds or consciousness could have crafted search schemes or aimed at targets. Only inanimate physicochemical reactions, interactions and phase changes existed prebiotically on earth in any naturalistic scenario. The environment had no goals and no sense of functional success. Steering toward biochemical and metabolic "success" was non-existent. In addition, steering and programming were logically impossible. Chance and necessity, mass and energy, cannot program. Programming requires purposeful choices at bona fide decision nodes. Naturalistic philosophy disallows choice with intent from the realm of possibility prior to the existence of organismal central nervous systems.

So, how did the first protocells get organized, programmed and processed?

Whatever the answer, it cannot be merely epistemological. Naturalism must be able to point to an ontological scenario. An objective, historical sequence of events had to be able to produce organization, programming and processing for naturalistic belief to be tenable. In addition, the epistemological hypothesis or model about these supposed objective events must be *scientifically plausible*. The scenario must sport a Universal Plausibility Metric value ξ of greater than 1.0 to be considered a scientifically plausible model (the Universal Plausibility Principle[80]). Finally, the proposed scenario must be free of logical fallacies. It must be rationally sound.

No such ontological sequence of events has ever been observed and demonstrated that could generate non-trivial functionality. No data has been forthcoming from life-origin literature (apart from artificial steering and investigator involvement in experimental design).

Since zero data exists showing spontaneous self-organization, programming and processing of that programming, the UPM (ξ) of any such naturalistic model is certainly well below 1.0, considered to constitute scientific falsification of that model.[80]

Finally, the notion of self-organization is logically untenable. Nothing can organize itself into existence. An effect cannot be its own cause.

Since all known life is cybernetic, ontological P & P, independent of earthly organismal central nervous systems, had to have organized the first life.

This fact alone is sufficient to reject the most fundamental axiom of metaphysical naturalism, that "Physical nature is sufficient" to explain life and reality. Physical nature is clearly not sufficient to explain the origin of life. This is not a metaphysical conclusion. This is a scientific conclusion. Can we go beyond this point *scientifically*, to conclude anything more? Probably not, except to point out that universally, so far, none of the

following phenomena have ever been observed to occur independent of agent involvement: valuation of functionality, organization, steering toward success, pursuit of utility, programming choices, computation, even the scientific method itself. None of these phenomena have ever been observed independent of agents. All of these are formalisms originating from the far side of The Cybernetic Cut, not the near side.
Materialism/physicalism/naturalism cannot escape the reality of nonphysical formalisms. Naturalistic *philosophy* cannot deny formalism's dominance and control over physicality both in physics and biology (the F > P Principle).[19,29]

3.28.9 We must first explain the *origin* of PI_o, and its processing, before addressing modification of existing PI_o.

We cannot possibly understand all of the ramifications of biological PI_o modification through mutation without first understanding PI_o's initial programming, its methodology of prescription, and its efficaciousness. How did initial PI_o get optimized sufficiently for life to begin?

Genetics constitutes freely resortable material symbols systems. Physical tokens (nucleotides) are selected using an arbitrary, conceptual block code methodology of three symbols for each codon. Each codon then represents an amino acid or stop instruction. Other layers of prescription of function are hidden in the supposed redundancy of the codon table,[253,254] proving that the supposedly redundant codons are not "degenerative."

How did a prebiotic environment generate the first symbol system?

How did an inanimate environment program *future* computational success?

How did nature generate the first coding and translating systems?

Epigenomics involves very real configurable switch settings. Life is just as dependent upon switching on and off genes as it is upon those genes themselves. These configurable switches can only be set to achieve precise regulation and pragmatic success in a formal choice-determined (CD) sense, not a physicodynamically-determined (PD) sense.

Biological information (BI) is a form of Functional Information (FI). Descriptive Information (DI) and Prescriptive Information (PI) are both subsets of FI. It is important to realize that Biological Information (BI) is PI rather than DI. In addition, it is important to realize that BI is Ontological PI (PI_o), not Epistemological PI (PI_e). The key to life is genomic and epigenomic prescription (PI_o), plus its processing. The two together are called P & P_o. Both prescription and its processing are formally mediated functions, not merely physicodynamic phenomena. They are formally controlled, not just constrained.

The PI_o of life has little in common with Shannon "information" (SI). Shannon "information" is actually nothing more than Shannon "Uncertainty." The only thing that both Shannon Uncertainty and PI_o have in common is "contingency." Contingency is the source of Uncertainty and Possibility. But Uncertainty and Possibility are not prescriptive of non-trivial function. FI is quite real, but Szostak's measurement of what he and Hazen *call* "functional information" is bogus. Their measurement attempts to use stochastic theory to quantify "a stipulated uncertainty." But, no uncertainty can be specifically stipulated. Upon its stipulation, it is no longer uncertain. Stipulation of what works requires at least Descriptive Information (DI) about which specific sequences will work, and which will not work. At that point, we are no longer talking about Uncertainty. We are talking about the relative "certainty" of what works. No measure exists for the Prescription and Processing of what works. We can use DI to reduce the overall uncertainty, but that reduced overall uncertainty is still not information. It remains Uncertainty, even though reduced. The only source of true information was the FI externally supplied that was used to reduce the uncertainty. True positive extrinsic FI has to be smuggled in through the back door to know which combinations work, and which don't. Without this provided FI, the Uncertainty cannot even be reduced.

Notice that the reduction of Uncertainty, known as mutual entropy, corresponds to a reduction of "possibility." We always appeal to "possibility" to try to explain, naturalistically, how life's Prescriptive Information (PI_o) could have spontaneously arisen. But maximum possibility requires maximum Shannon Uncertainty. Maximum Shannon Uncertainty is the antithesis of mutual entropy. If mutual entropy is information, why does an increase in mutual entropy (a reduction of Uncertainty) simultaneously reduce possibility? Current concepts of "information" in naturalistic science are thoroughly confused and misleading.

Stochastic Theory cannot measure Functional Information (FI) (See Section 2.16). Stochastic Theory is blind to meaning and function, as Shannon himself pointed out in his very first paper.[67] The logical fallacy of a "category error" is committed when we try to combine the certainty of what works with the uncertainty measured by "bits." We might be able to measure the percentage of stochastic ensembles that work using extraneously-introduced information to get a measure of Functional Sequence complexity (FSC).[6,72,74] But, we cannot legitimately measure Functional Information with any modification of Stochastic Theory. Bona fide information is a function of Decision Theory, not Stochastic Theory. Purposeful decisions, and their relative worth to computational programming, cannot be measured with statistics.

The simplest binary node is a fork in the road. One can flip a coin to decide which fork to take. But a coin flip would not be likely to get you to your destination any faster. No computational program has ever been written by coin flips or random number generators.

One cannot use "chance" to compute anything. Computation is formal and choice-based (choice-contingent rather than chance-contingent.) But random events can be described and predicted with stochastic mathematical theory, including weighted means with partial causation and partial randomness mixed. But chance cannot cause anything. Chance is a purely descriptive mathematical construct, not a cause of physical effects. We might be able to describe quantum events statistically. But that statistical description does not cause or explain any quantum effect. The same is true in the macroscopic world. Chance is not a cause of *any* effect. More importantly, chance has never once been documented to produce a single non-trivial, sophisticated function of any kind.

How easily and subconsciously we sneak in investigator involvement without realizing it. Then we attribute those human steering or computing choices to purely physical interactions. This happens all the time with supposed evolutionary experiments where the experimenter chooses which iteration to pursue next based on human desire *to engineer* a functional ribozyme, for example. It's like Dawkins programming his computer to "evolve" his desired phrase.[354] At every step, his program purposefully Selects *FOR* a predetermined target phrase. Most of the half-written phrases along the way have no meaning or selectable function. The supposedly "evolutionary" program is instructed by programming choices to pursue the goal of a certain phrase. But real evolution has no such goal.[539]

There is an underlying mathematical, formal and rational nature to functional physicality that is being progressively discovered, the same as when a new mathematical physical law is realized. We see it in the periodic table, in fibonacci numbers, and genomic programming.

Science is about repeated observation and prediction fulfillment. Chance doesn't hack it as an explanation for the clearly observed genetics, epigenetics and genomics that instruct, control and regulate even the simplest cells.

It has not been documented at all that the complex eye evolved from the simplest forms of eye step by infinitesimal step. This has only been imagined with a great deal of wish fulfillment and blind belief. Imagination is not all bad. It has led to many important discoveries. But we do have to distinguish between models/theories and repeated observation/prediction fulfillments.

The statement that neither chance nor necessity can program is an absolute logical imperative, not an inference that could be overturned by new data around the next corner. Programming of any kind derives only from Decision Theory at bona fide decision nodes, not through descriptive Stochastic Theory. It is a *logical impossibility* for either chance or necessity to select representational symbols from an alphabet of symbols in a symbol system, or to program dynamically-inert configurable switches so as to integrate circuits. Yes/No, On/Off, Open/Closed purposeful choices must be made in pursuit of future computational halting, before that computational success ever exists to be phenotypically and secondarily selected by the environment.

One of four options must be selected at each locus in the string of a polynucleotide prescriptive message. Each nucleotide "choice" represents elimination of two rather than one bit of Shannon Uncertainty. The amount of PI required for RNA or DNA instruction is far greater than for a computer program. DNA affords the greatest information density known to computing. You may have noticed in very recent literature that DNA is now being used in cybernetics to record libraries of true information (as opposed to mere Shannon Uncertainty).

The passive voice used here, and the specific wordage of other phrases, connotes agent involvement without proponents of evolutionary algorithms realizing it. We think we are offering a naturalistic model, but we are not.

Of course it is now evident from ENCODE research that virtually none of DNA is junk. 98% is highly instructive (prescriptive) switching instructions and assemblers of bits and pieces of Turing tape from all over the genome into overlapping genes, complementary strands being read in the opposite direction, etc. Even the number of repeats is prescriptive. *Mycoplasma genitalium* is probably the simplest known organism. Yet, the multi-layered and multi-dimensional conceptual and prescriptive nature of its genomics and epigenomics puts to shame the finest mainframe computer systems in the world.

Necessity is the only other option to consider within naturalistic presuppositions if chance can't explain genomic and epigenomic programming (e.g., control and regulation). Maybe some yet-to-be discovered law wrote DNA programming instructions.

Clay adsorption polymerizes polynucleotides spontaneously without any enzymes. This is extremely inviting for life-origin theory. Enzymes could not have existed in a prebiotic environment. Only short peptides could have existed. The problem with clay adsorption polymerization, however, is that such prebiotic self-assemblies are devoid of information potential. They have no Shannon uncertainty, because adenine is always "selected" (in reality, constrained, not selected). Polynucleotides would all be polyadenosines, for example. A polyadenosine is DNA or RNA with only one base, adenine, at every locus in the string of nucleotides.

If some cause-and-effect Law had dictated the same selection at each decision node (the same nucleotide selection from among 4 options), every programming "choice" would be the same, by Law! The program would have consisted of all 0's. Or, the program would have consisted of all 1's. Such a "program" could not have computed anything.

RNA chemists know that polyadenosines adsorbed onto montmorillonite clay are useless for life. They have no sequencing to instruct folding into ribozymes. And they cannot contain any information because they contain no Shannon uncertainty. Law forces redundant self-ordering. Necessity constrains and militates use of the same nucleotide, rather than any of the other three, at every locus in the string. All the bases are the same in the string. A polyadenosine has zero information content, because it has no freedom at its "decision nodes." This, in reality, means it *has no* decision nodes!

Some laws are probabilistic, such as in quantum mechanics. Note that these "laws" are purely statistical and descriptive, not prescriptive or causative of physical effects. Stochastic Theory cannot replace Decision Theory when it comes to explaining the obvious choice-contingency necessary to generate the obvious cybernetics of life. We investigate and freely acknowledge the reality of cybernetics in molecular biology. It is only our naturalistic *metaphysical* commitments that preclude our admission that cybernetics is impossible without choice-contingency. No decision node can be programmed without a bona fide decision! Decisions require choice-contingency, not chance-contingency or law.

4. Summary

The most plaguing problem of life-origin science is failure to explain the origin of objective, ontological, non-epistemological Prescription & its Processing (P & P) in an initial prebiotic environment.

To draw the loose ends together, we must review the basics.

Contingency means that events can happen in multiple ways despite cause-and-effect determinism. Contingency requires freedom from law-like constraints. Two types of contingency exist: Chance-Contingency and Choice-Contingency. This is called the Universal Contingency Dichotomy (UCD). (See Section 2.7)

The Prescription Principle states that Prescription and its Processing (P & P) require purposeful Choice-Contingency to generate. Purposeful choices are caused by Choice Determinism (CD), not Physicodynamic Determinism (PD).[2,6,7,21,22,31,39,41,42,62,82] Physicodynamics refers to physicality and mass/energy interactions. Another name for CD is Choice-Contingent Causation and Control (CCCC).

P & P alone *produces* true organization, not just redundant order. P & P alone also produces sophisticated function. What matters to life-origin science is the actual *causation* of primordial, sophisticated formal function. How did the first computation and integration of component parts arise in an inanimate environment? Mere information *about* primordial function in the minds of Johnny-come-lately humans is not the issue, and does not provide the cause of P & P in a prebiotic environment.

Controls are needed to steer entities and events toward organization and sophisticated function. Constraints and physical laws cannot do this. Constraints and laws are blind to utility.

Section 4: Summary

Decision theory requires freedom from fixed, forced law. Stochastic, random events cannot cause CD, either. Decision Theory requires purposeful choices at true "decision nodes," not probabilistic events at mere bifurcation points. Only when purposeful choices are made at forks in the road do those bifurcation points become bona fide "decision nodes."

Rats improve the efficiency of their escape from mazes only by exploring wiser *choices*. Coin flips (Chance-Contingency) at forks in the road do not improve the efficiency of a journey. Wise, purposeful Choice-Contingency can.

Formal means choice-contingent, or purposefully choice-determined. Formalisms are not dependent upon the laws of physics and chemistry.[19] But, they do not violate those laws, either. The most basic binary configurable switch first provides "contingency," because it can be set to either position, open or closed, despite cause-and-effect physical-law determinism.[22,70,79] But upon the setting of that configurable switch, it acquires the additional characteristic of having been Choice-Determined (CD) by formal CCCC.[41,42] The choice is both purposeful and free from physicodynamic determinism. Energy may be required to actually set the switch, but the choice of which direction to push the knob, or slide the physical trap door (open or closed), is purely formal.[19]

All formalisms arise from the far side of The Cybernetic Cut (See Section 2.13). The Cybernetic Cut is perhaps the most fundamental divide of reality. The law-like orderliness of nature, along with the seeming chance-contingency of heat agitation and statistical quantum reality, lie on one side of the divide. Choice-contingency lies on the other. Choice-contingency is the ability to choose with intent what aspects of being will be preferred, pursued, selected, rearranged, integrated, organized, preserved, and used. Chance and necessity cannot generate choice-contingency.

Most of what is really interesting in presumed objective reality is formal. The epistemic cut of Howard Pattee has to do only with human knowledge.[11-13,16,17,53,321,540] The Cybernetic Cut addresses ontological (objectively existent) Prescription and its Processing, and their origin in a prebiotic environment devoid of any knowledge or epistemology.

The Cybernetic Cut can only be traversed through nonphysical, formal, purposeful, decision-node choice-commitments. The only known route from the far side of the Cybernetic Cut to the near side is the Configurable Switch Bridge (CS Bridge) (See Section 2.14). This is a one-way bridge over the Great Ravine. Traffic can flow only from the far formal side to the near physical side of The Cybernetic Cut.

Configurable switches must be set with intent, purpose and goal if computational success is to be realized. Such choices are instantiated into physicality using dynamically-inert configurable switch-settings. Material Symbol Systems (MSS)[44,45,522] can also be used to instantiate formal function into physicality. Physical symbol vehicles (tokens) must be purposefully chosen from among an alphabet of tokens to create meaning and function. The integration of component parts into a holistic device (e.g., a machine) is another means of instantiating formalism into physicality.

Physicodynamics, on the near side of the Cybernetic Cut, possesses no ability to choose with intent at decision nodes. Physicodynamics cannot assign meaning to symbols, ascribe value to functionality, or pursue utility. Info-dynamics (trying to reduce Prescriptive Information [PI] solely to physicality) is logically impossible. Even Shannon Uncertainty Theory is formal, not physical. The notion of info-dynamics provides no mechanism for the spontaneous generation of Prescriptive Information (PI) or its processing. This includes the genetic instructions required for metabolic organization and life.

Algorithmic optimization also requires traversing the Cybernetic Cut across the CS Bridge, from the far side to the near side of the great ravine. Physicalism provides no plausible explanation for, and no empirical evidence of, unaided "self-organization,"[62] despite use of the term in hundreds of published papers.

The necessity of traversing the Cybernetic Cut in order to instantiate functional controls over physicality is a fully falsifiable principle. The observation of a single case of non-trivial spontaneous computation, independent of agent steering, would suffice. Falsification experiments would have to be free of hidden artificial selection. Iterations cannot be steered by experimenters as we see in SELEX experiments of ribozyme engineering.[541-543] The experimenter chooses iterative products and pursues the use of those artificial choices. The outcomes are in direct violation of evolution theory.

So-called "evolutionary algorithms" are invariably examples of "directed evolution." Both of these last two terms are self-contradictory nonsense terms. If a process is directed, it is not evolutionary. If the process is evolutionary, it is not directed. Algorithmic optimization is invariably steered toward the goal of ideal utility by programmer choices. Evolution has no such goal.[155] Traversing the Cybernetic Cut is the necessary and sufficient condition for generating any formal control system's governance of mass/energy interactions.

Principles of science must not only be falsifiable, they must provide a historical meta-narrative and explanation across a wide range of phenomena. In addition, they should foster verifiable predictions in unrelated fields. What predictions does the Cybernetic Cut afford?

1) No non-trivial computational function will ever spontaneously arise in any physicodynamic medium or environment independent of formal intervention and controls.

2) No sophisticated, algorithmic optimization will spontaneously proceed upon removal of hidden experimenter choices and steering of iterations (illegitimate "investigator involvement" in experimental design).

3) No non-trivial functional controls of physical phenomena will be realized independent of the programming of dynamically-inert (dynamically-incoherent) configurable switch-settings that alone instantiate formal agent choices into physical reality.

A single verifiable occurrence of any one of these three null-hypothesis predictions would falsify the Cybernetic Cut.

Organization arises only out of Choice Determinism (CD), not Physicodynamic Determinism (PD). While physicodynamics can spontaneously self-order to form dissipative structures, only the choice-contingency of formalism can organize integrated functional systems. Organization is formal, not physical. Prigogine's dissipative structures in Chaos Theory are self-ordered, not self-organized.[62] "Self-ordering" in nature is not "organization".[62] Chaos theory does not and cannot deal with organization. Chaos theory only deals with self-ordering.

"Self-organization" is an oxymoronic self-contradiction. Nothing can organize itself into existence. An effect cannot be its own cause.[544] The notion is logically impossible even before we point out the complete lack of observational support and prediction fulfillments.

Sustained Functional Systems (SFS) cannot be generated by mere complexity, phase changes, "buttons and strings," or imagined interfaces between chance and necessity (See Section 2.4.4 and 3.11). The proposed capabilities of the "edge of chaos" are pure science fiction.

Increasing Entropy is better thought of as increasing disorganization rather than increasing disorder. (See Section 3.11.5) Formal organization and pursuit of function are required to locally and temporarily circumvent a relentless increase in entropy.

Useful work is intuitive, functional work. The physics definition of work is completely unrelated to the intuitive, every-day, semantic connotation of "useful work" that prevails in human minds. Physicodynamics and physicochemical interactions/reactions have no perception of formal, abstract concepts such as "usefulness." Physicality, apart from agent sentience, is utterly blind and indifferent to notions of "utility" and "value." Physics and chemistry play no direct role in generating "functionality." The laws of physics and chemistry *can be used by agents* to achieve utility. But that involves steering and control, not just constraint and law. The value of "functionality" is entertained only within the minds of personal agents. Inanimate nature values and pursues no such thing.

When we talk about protometabolism in a protocell, we are talking about needed, *useful work* having to be accomplished.[331] Even the simplest metabolism doesn't just happen. Conceptually complex biochemical pathways and cycles must be directed and steered toward creative metabolic success. In this formal process, energy and the ability to use it are both required. The tendency here is to glibly say, "Usable energy is required." What is required is sophisticated *formal systems* that are able to harness otherwise wasted solar energy, for example, transduce, store, and call it up when usable energy is needed. Sunlight is not "usable" energy. Sunlight is wasted energy from a burning star. The only thing that makes it "useful" are formally controlled Sustained Functional Systems (SFS), such as chloroplasts. The genomes and epigenomes of plant cells Prescribe and Process chloroplasts into existence. They don't just happen. There are, of course, simple "light-induced" chemical reactions. But, these are never formally organized by abiotic nature into holistic metabolic schemes.

Machines are a type of SFS. Machines continue doing useful work for extended periods of time. They are not dissipative like Prigogine's momentary, self-ordered states that do nothing useful, and that usually only destroy organization (e.g., tornadoes and hurricanes).

Not all machines were designed and engineered by humans. Millions of subcellular machines are hard at work every nanosecond. Many of these machines were needed from day one in any primordial living cell. How did all these machines just happen to occur at the same place and time? What made them all cooperate in achieving the formal goal of homeostatic metabolism and life?

The prebiotic environment did not perceive or consider formal concepts such as pragmatic goals. One chemical reaction was just as good as any other. No physical interaction or phase change was preferred over any other, except for thermodynamic tendencies in accord with the Second Law. This most basic tendency of physicality does not produce physical states far from equilibrium. Only Maxwell's demon's purposeful opening and closing of the trap door between compartments produces states far from equilibrium. Maxwell's demon is an *agent*.

Any pragmatist should strongly object to naturalists trying to use the physics definition of work to refer to intuitive "useful work." The physics definition of "work" is utterly blind to utility, and cannot possibly address or explain the "useful work" required by cellular metabolism.

Two kinds of selection exist, as defined by the Universal Selection Dichotomy (USD):[21,22,29,31,42] Selection *FROM AMONG* vs. Selection *FOR (in pursuit of)*. (See Section 3.3)

Cybernetics (controlled processes) requires Selection *FOR (in pursuit of)* at bona fide decision nodes, not mere coin flips at bifurcation points (forks in the road). No programmed computation can be achieved with mere Selection *FROM AMONG*.

In artificial cybernetics, Selection *FOR (in pursuit of)* is what we recognize as purposeful choices made in an attempt to reach some destination, goal, or to accomplish a utilitarian task. Selection *FOR (in pursuit of)* is what transpires when we design, create or engineer something into useful existence. Selection *FOR (in pursuit of)* is required to generate processes and manufacture a conceptually complex product. Programming computer software is a classic example of Selection *FOR (in pursuit of)* the not-yet-existent function of computational halting success. Computational halting is a goal, not yet a reality, when the prescriptive choices at true decision nodes have to be made. We can't even be certain that the finished program will successfully compute (the famous "halting problem"). Yet the programming choices have to be made and recorded (instantiated) into physical configurable switch-settings or Material Symbol Systems (MSMs).[44,45,522]

Planning a meeting's agenda, deciding what to teach a class, writing a book outline, designing and engineering a new invention, bridge building, all involve purposeful choices made in pursuit of desired, not-yet-existent functions. Any kind of sophisticated, integrated function requires Selection *FOR (in pursuit of)*. Selection *FROM AMONG (from among existing options)* cannot possibly program, prescribe, instruct, or manufacture sophisticated new functional entities. Only Selection *FOR (in pursuit of)* alone can formally organize disparate components into a holistic complex machine or processes.

When Maxwell's demon selects a trap door setting "in order to concentrate the faster moving, hotter atoms on one side of the partition," he is Selecting *FOR (in pursuit of)* potential function (a usable energy potential). Other Selections *FOR (in pursuit of)* would be necessary to make use of that energy potential (e.g., constructing a primitive heat engine). This is quite different from natural selection, where the environment "Selects" the fittest *FROM AMONG* already-programmed, already-living phenotypic organisms.

Neither molecular evolution nor post-biotic evolution can Select *FOR (in pursuit of)*. Natural selection (NS) never exceeds Selection *FROM AMONG*. NS has no goal. NS never selects for *potential* function, especially not at the genetic prescriptive level of symbol system and Hamming block code (triplet codon) use. This fact is known as the GS Principle of Biology.[71,83] (See Section 3.4.3) Genomic and epigenomic programming required Selection *FOR (in pursuit of) potential* utility, in advance of any naturally selectable fitness.

Evolution is nothing more than the differential survival and reproduction of the fittest already-programmed, already-living phenotypic organisms. Evolution does not work at the molecular/genetic programming level (also part of The GS Principle).[71,83] Natural selection is eliminative only. Evolution cannot and does not program sophisticated new computational function and utility.

Evolution has no desires, values, pursuits or goals. Evolution, therefore, is a non-theory when it comes to abiogenesis and the programming of initial Prescriptive Information (PI). No "pressure" whatever exists in evolution, except the pressure of Freudian wish-fulfillment. Any pressure is only imagined, in order to make our model "work for us." The price that is paid is the loss of correspondence of that imagination with objective reality. Neither molecular evolution nor *post*-biotic evolution can Select *FOR (in pursuit of)*. Yet the latter is the essence of programming. Life is clearly programmed. That programming could not possibly have evolved into existence.

No living organisms existed in a prebiotic environment from which to select. Selection *FROM AMONG* living organisms, therefore, was nonexistent. Neither molecular nor post-biotic evolution could have participated in Selection *FOR (in pursuit of) potential* function. This leaves outstanding the question of how *any* programming or processing could have arisen in a prebiotic environment.

All known life is cybernetic. Cybernetic means controlled, not just constrained.[61] Life is tightly regulated with many positive and negative feedback mechanisms. Without this exquisite fine-tuning and management, homeostatic metabolism would be impossible in highly variable environments.

Fittest organisms only exist because of superior programming, editing, and processing of Turing-tape-like instructions. Configurable switches must be set by Choice Determinism, not by chance or necessity, if computational success is to be realized.

With the exception of Ontological Prescriptive Information (PI_o), information is usually epistemological knowledge *about* reality. Science is an epistemological system. The goal of science has always been to bring human knowledge into closer approximation, or progressive correspondence with, the way things actually *are* in a presumed objective reality *being*. This requires ruthless self-honesty in evaluating popular theories, and even prevailing paradigms. Selection "pressure" does not cut the mustard as a scientific theory of origin for Prescription and its Processing (P & P).

The emergence of Selection *FOR (in pursuit of)* is essentially the same problem as the problem of emergence of consciousness. Programing using a symbol system like DNA codons and their formal translation table to amino acids is fundamentally formal, not physical. Choice Contingent Causation and Control (CCCC) and Choice Determinism (CD)[19,29,31,42] cannot be explained by mere physicodynamic and physicochemical determinism. The origin of subcellular CCCC and CD cannot be explained by evolution, either, where selection is limited to Selection *FROM AMONG* already-existing, already-living, already-programmed phenotypic organisms, rather than Selection *FOR* (in pursuit of) (See Section 3.3).

Trying to explain how PI_o emerged from chance and necessity is not an anthropological, psychological, or even a biological problem. It is a logical impossibility. The phenomenon of choice-contingency did not start with the central nervous systems of metazoans. It started with the very first protocell. Naturalistic science will not be able to escape the simple objective fact that the first cell required Choice Determinism to be programmed into existence.[5,26,27,315] The generation of Prescription and its Processing (P & P) constitutes formalisms. Formalisms require agency to make new purposeful choices. Choice-contingency is the essence of any formalism.

Why are all these considerations so important to life-origin science, and to molecular biological information theory? The answer is that *the subject of these studies is Ontological Prescription, not epistemological information.* We are not studying our own consciousness and epistemology. We are studying an aspect of stand-alone objective reality that produces, controls, governs and regulates life at the subcellular, cellular and intercellular level. This takes place independent of human uncertainty, and knowledge. The cause of P & P is ontologically "stand-alone." Prescription and its Processing (P & P) cannot be reduced to the epistemology of higher vertebrates. Vertebrates did not exist three billion years ago. PI_o is far deeper and wider than human epistemology. PI_o extends into areas far beyond human uncertainty or concepts can address.

PI_o generated the physical brains which materialism claims secrete knowledge. We would do better concentrating on *what secreted physical brains* that supposedly secrete minds! Embryological development of central nervous systems will never be explained by mere mass/energy interactions. Development is programmed. Development is formal rather than physical, not that development doesn't *use* physical tokens, matrices and laws to accomplish its P & P.

We want to examine various scientific models of spontaneous protocell formation. Could such protocells possibly have self-assembled in the absence of P & P? If so, how true to empirical life are these models? Are theorized protocells really alive?

Life requires exceedingly tight management. The simplest life forms known cannot exist without highly integrated, cooperative, metabolic pathways and cycles. Biochemical pathways must be steered toward useful endpoints. Subcellular reactions must be highly coordinated. A high degree of energy efficiency is observed throughout all living processes. Metabolism pursues and maintains the undeniable *goals* of being alive, and staying alive. Not only does nature not have goals, neither does evolution have any goals.

"Combinatorial possibility" is nothing more than Shannon uncertainty. Combinatorial possibility doesn't prescribe anything. It cannot be equated with "Prescription of Function (PoF)." With post-biotic evolution, we attribute the prescription of new function to "variation" of existing programming, which is then only secondarily favored phenotypically. In the case of abiogenesis, no prior programming existed to "vary." In a naturalistic world, any prebiotic rise in function or heritable prescription had to have occurred spontaneously. Zero empirical evidence or logic exists to support blind belief in spontaneous programming or self-organization in a prebiotic environment. The self-ordering of chaos theory has nothing to do with pragmatic organization. Neither chance nor necessity can formally organize. Mass and energy cannot make programming decisions or pursue utility.

"Variation" can only modify existing semantic, functional, Prescriptive Information (PI), not program it. Any modification of existing PI by mere variation is no more helpful than typographical errors are to any instruction manual.

Order and randomness are at opposite ends of the same Complexity bidirectional vector. They are both in the same dimension of complexity, only at opposite extremes. Randomness is maximum complexity, and maximum disorder. Maximum complexity has nothing to do with function or organization.

Maximum order represents the opposite of randomness and the opposite of maximum complexity/maximum Shannon uncertainty. But, neither order nor randomness (maximum complexity) can formally organize anything! Organization is in a completely different dimension from the order vs. randomness/maximum complexity/maximum uncertainty dimension (See **Figure 4** in Section 3.11.5). As physicality moves toward heat death, organization is always lost. Order is usually lost also, but not always. Entropy can increase with self-ordering crystallization, for example.

The obvious organization and goals of life processes present a real problem for naturalistic science. Evolution has no goals. Naturalistic science eschews acknowledging the reality of any goal in nature. To admit to goal is seen as giving up ground to teleology. Teleology is seen as pure metaphysics, if not outright religion. Metaphysics means literally, "Beyond physics," or "Beyond naturalistic science." Admitting teleology into naturalistic science "would surely invite in 'the crazies.'" Thus, we will go to just about any lengths to avoid admitting the existence of any "purpose," "goal" or "choice" in nature. In fact, we will flat-out deny their existence in order to maintain a purely physicalistic philosophic perspective of reality. Living systems can only "appear" to have purpose. They cannot actually *have* real purpose or a goal.

The first problem with this is that declaring that "Physicality is all there is" *is also pure metaphysics.* Somehow, we remain blind to the realization that the pre-assumption of "no goal or purpose" is every bit as much of a purely metaphysical presupposition as its opposite. No consideration whatever is given to the very real logical possibility that "the appearance of" purpose could, in fact, actually *be* purpose. That possibility is utterly precluded by prior commitment to absolutist naturalistic philosophic dogma. The open-mindedness for which we claim such rationalistic superiority goes out the window. "Physicality did it, period!" or "Don't bother me with the empirical evidence of genetics or ENCODE findings.[250-252,509] My mind is made up!" and "Physicality is all there is, ever was, or ever will be." Nowhere in this pontification is an answer to the question asked by a number of Nobel laureates about "The unreasonable effectiveness of non-physical, formal mathematics in the physical sciences."[300-303] Nowhere in the dogma of materialism, the backbone of "naturalistic science" *philosophy*, is any acknowledgement that scientific method, and science itself, are completely nonmaterial, formal and "non-natural."

We can no longer rationally deny, or successfully obfuscate, the fact of purpose and goal in subcellular physiology. In naturalistic science, we eschew any hint of teleology. We use instead the euphemism, "teleonomy" in an effort to naturalize any concept of teleology. The problem is this: just renaming "teleology" to "teleonomy" does not solve the problem of explaining how chance and necessity, mass and energy generated such exquisite organizational concepts, true controls, and computationally successful programming. Both teleology and teleonomy must still be *steered* toward the goal of sophisticated utility. Without steering, the goal is empirically never reached. Despite all of its evolutionary smoke screens, naturalistic science knows of no sophisticated function of any kind that was not purposefully steered or programmed into existence. Purposeful choices had to have been made to direct events toward the goal of utility. Any argument to the contrary is just blindly believed, against all rational and empirical evidence.

Mere faith in "emergence" does not provide a plausible scientific mechanism for how stochastic ensembles of RNA analogs simultaneously sequenced themselves (all with identical chemical bonds) into Prescriptive Informational Programming Strings (PIPS).

Metabolism depends entirely upon purposeful steering, control and regulation. Holistic metabolic success is impossible without formal controls. Life depends upon such controls, not just constraints.[61] Life also requires organization, not just redundant order or patterns.[62] Life doesn't just send "signals" to itself. It sends meaningful, functional, semantic "messages." Circuits have to be integrated. Even the simplest bacteria actually compute their metabolic success. The most primitive cells are highly programmed.[5,26-29,347]

Survival of the fittest has nothing whatever to do with programming and prescribing life into existence. Evolution, molecular evolution included, is a worthless non-theory when it comes to life-origin models.[29] Both pre- and post-biotic evolution involves only Selection FROM AMONG already-existing living organisms. No naturalistic mechanism exists for programming sophisticated new function.

Life-origin research should concentrate primarily on the emergence of Prescription and its Processing (P & P) in a prebiotic environment. The most fundamental question of abiogenesis is not whether water or methane exists on a planet or moon. The most fundamental question is not whether an analog pre-RNA World existed. It is, **"How did an inanimate environment prescribe and process organization, control and regulation of protocellular metabolism, and eventually primordial life?"**

To study abiogenesis and life's controls and regulation systems requires that we realize that the subject of our study is Decision Theory, not Stochastic Theory.

The next question becomes whether natural selection can select monomers at the polymerization level. The answer is NO! (See Section 3.4, and The GS Principle.[71,83])

Naturalistic life-origin science is faced with a real dilemma. What makes this problem so acute is that we cannot explain Selection *FOR (in pursuit of)* from a naturalistic perspective. We observe this phenomenon repeatedly every day, not just in artificial arenas, but within every cell. Yet, we disallow Selection *FOR (in pursuit of)* for purely metaphysical reasons, *a priori*. We brought with us *to* science certain unproven presuppositions. Not only that, we incorporated these purely philosophic pre-assumptions into our very definition of science, despite knowing perfectly well that the vast majority of the greatest scientists of history did *not* find it necessary to incorporate naturalistic philosophic presuppositions into *their* definition of science. Scientists such as Kepler, Newton and Faraday did quite well elucidating the most crucial principles and laws of science despite being highly critical of naturalistic faith.

The most fundamental Principle of science is *not* the First or Second Law of Thermodynamics. It is the Formalism > Physicality (F > P) Principle.[19,29] The F > P Principle states that "Formalism not only describes, but preceded, prescribed, organized, and continues to govern and predict

Section 4: Summary

Physicality." The F > P Principle is an axiom that defines the ontological primacy of formalism in a presumed objective reality that transcends both human epistemology, our sensation of physicality, and physicality itself.

Science is about repeated observation, prediction fulfillment, double-blind experimentation that eliminates investigator involvement that might affect the results. Science is also dependent upon sound logic.

We know immediately, therefore, that any worldview that would attempt to reduce reality to nothing more than chance and necessity of physical nature is bogus. Such a worldview defines a theoretical perimeter outside of which abundant empirical reality clearly exists. "Chance and necessity is all there is!" is worse than fanatical religious dogma. It is a patently erroneous pre-assumption. To the true believer in this dogma, we need to ask not only, "SEZ WHO?" but, "How was that scientifically determined?" We can immediately point to an abundance of observational data that proves that subcellular programming exists, that programming cannot possibly be produced by chance and necessity, and that this abundant programming is outside the perimeter set by materialism / physicalism / naturalism. In other words, we can declare with confidence that these worldviews are not only utterly inadequate, but are clearly falsifiable as scientific models. None of these philosophic metanarratives can contain all of the pieces of the puzzle that we have in hand. Reality is clearly bigger than what naturalism *says* reality is.

All logic, even mathematical logic, flows from starting unproven axioms. Abundant evidence exists in the history of science of how "good ole boy" hierarchies of scientific control have impeded scientific progress by presupposing erroneous axioms. *The majority opinion, or the perspective of the ones in control, however, is not always the presupposition that best corresponds to objective reality.*

Just as the formal mathematical equations of physics could not have arisen from physicality, the P & P of life could not have arisen from physicality. The current axiomatic metaphysical system upon which naturalistic science is based—a very dogmatic religious physicalism—does not correspond to/with repeated observations of formal genetic, epigenetic, and multi-layered/multi-dimensional cybernetic management of life.[253,455,466,487,507,545-547] Any argument for physicalism is inconsistent with its own premise, and is rationally self-contradictory. "Both the governing mathematical laws of physics and the computational programing of life dictate that nonphysical formalism precedes, organizes, controls, manages and regulates the physical cosmos (The Formalism > Physicality [F > P] Principle).[19] Ontological *being* (objective reality) is fundamentally rational, not irrational. Explosions have never been observed to produce mathematical rationality, programming or processing

Nature could not have produced nature. This is not just true with biology. Physicality could not have produced physicality. The laws of physics themselves are non-physical. They are formalisms. All formalisms are choice and rule-based. Constraints and physical laws cannot generate any formalism. The laws of physics are themselves mathematical and logical. At least a third category of reality is required to explain all of the empirical data: the category of Choice Determinism (CD).

Programming of anything, along with mathematical computation, is logically impossible without making purposeful choices at bona fide decision nodes. If our worldview disallows the reality of purposeful choices, we not only can't explain computer programming, we can't explain genomics/epigenomic programming, either. It is a logical impossibility for Chance and/or Necessity to have programmed DNA instructions. Naturalism, with its physicalistic/materialistic base, presents a false dichotomy (Chance OR Necessity). Many scientists have regrettably incorporated this sorry philosophy into their very definition of science, without realizing the utter illogic of their ways.

No hope exists of a naturalistic model providing an explanation for formal organization. We have no hypothesis whatever for how the motor protein kinesin walks along microtubules with two legs transporting other molecules to their destination.

We have no clue how ribosomes formed. They are required to manufacture proteins, yet ribosomes themselves require large numbers of proteins plus rRNAs. Scores of chicken-and-egg paradoxes now exist in life-origin science. Every last one of these chicken-and-egg paradoxes immediately resolves when we stop interfering with open-minded science. We need to simply acknowledge the obvious use of Selection *FOR (in pursuit of)* by life.

We have no naturalistic models whatever for the origin of Hamming block-coding used to reduce noise pollution in life's Shannon channels. Wong's,[548-554] Di Giulio's,[555-568] and Guimaraes'[569-576] commendable efforts to explain code origin fail to satisfy The Continuity Principle required of life-origin models.[577]

The Continuity Principle of life-origin science requires that proposed alternative models of earliest life connect smoothly with current life. How did stochastic ensembles of RNA analogs become DNA redundancy block-codes (codons) needed to prescribe each amino acid? How did codon language arise? How did it get conceptually translated into polyamino acid and protein language? The current codon table had to be in place and functional very early on. Recent findings elucidating the importance of supposedly redundant codons to Translational Pausing (TP) and protein folding make continuity all the harder to achieve.[253,254]

At the same time, these rigidly-bound biopolymers would have had to anticipate and instruct the correct minimum-free-energy folding structure needed to create auto-catalytic ribozymes. Ribozymes have many other needed metabolic and highly-specific binding functions. The same kind of "foreknowledge" was needed by nucleic acid biopolymers to instruct all of these superimposed functions. None of the needed protein molecular machines could have been genetically prescribed without knowing in advance what minimum-free-energy folding would be needed when each primary polyamino acid structure was sequenced. These functions all require Selection *FOR (in pursuit of) at the time of sequencing DNA monomers with rigid chemical bonds.* They require anticipation of what effects a certain sequence of monomers will generate before those monomers are ever polymerized.

Error-correcting mechanisms cannot possibly come into existence without Selection *FOR (in pursuit of)*. We cannot explain formal organization, the birth of bona fide controls, regulation, development, cell physiological management, integration of component parts, multi-step cooperative metabolic schemes, epigenetic switching, and multi-layered, multi-dimensional Ontological Prescriptive Information (PI_o) while trying to deny the obvious reality of Selection *FOR (in pursuit of)* (See the Universal Selection Dichotomy, Section 3.3).

Metabolism may use physical molecules, but metabolism is fundamentally abstract, conceptual and formal. Metabolism was in operation long before any central nervous systems existed to learn *about* metabolism epistemologically.

Epigenetics involves configurable switch-settings that turn genes on and off. But these highly functional switch-settings (e.g., methylation of cytosines) were *not* selections *FROM AMONG*. They were switch-settings that had to have been made *FOR (in pursuit of)* development, metamorphosis, and programming of optimal regulation prior to the existence of any of these phenotypic functions and organisms.

Denial of programming-type selection altogether is simply not tenable. Integrative, holistic, metabolic schemes are far too abstract, conceptual, interdependent, holistic and statistically prohibitive without Selection *FOR* (in pursuit of).

A favorite rationalization is, "There must be some yet-to-be discovered law that will explain programming." But, that notion is also ludicrous (See Section 2.7-2.8 and 2.13-2.15). No law can program anything. Laws, by definition, constrain physical interactions into the same redundant pattern every time. Laws describe very high probability events. We value laws because laws eliminate contingent behavior and combinatorial uncertainty.

Programming requires Choice-Contingency which in turn is characterized by *freedom* from fixed, forced law. Any hoped-for, "yet-to-be discovered new law of info-genesis," therefore, is a logical impossibility. Such a law would only preclude contingency of both kinds. First, Shannon Uncertainty would be extremely low. We forget that Shannon combinatorial uncertainty is the first requirement of info-genesis. Multiple real possibilities must exist for info-genesis and P & P to be possible. Second, any hoped-for new law would only preclude the freedom *to choose* from among real options. Programming would be impossible. No yet-to-be discovered law will ever be able to explain the phenomenon of computational programming and algorithmic optimization of formal function. Law is poison to formal organization and cybernetic regulation. Prescriptive Information can only be generated by Decision Theory, not Stochastic Theory. No decisions can be made if law forces events to unfold the same way every time (if law eliminates possibilities from which to choose).

The first protocell and primordial cells could not possibly have come into existence without formal controls at the subcellular level. Any happenstantial progress would have been extremely limited in any Metabolism First model. That progress would have needed to be retained. Re-inventing the wheel with each new pseudo-cell formation would not have worked. Some heritable system would have been required very early on. That means some Material Symbol System (MSS) that could prescribe phenotypic success, and that could mutate somewhat independent of that phenotype, would have been needed almost from the very beginning.

These physical symbol vehicles (tokens) would have had to have been selected and sequenced in advance of any computational or physical structural metabolic utility. Ribonucleotide sequencing in some RNA analog, for example, would have been rigidly bound chemically prior to the realization of any integrative bio-function. Ribonucleotides would have had to be Selected *FOR (in pursuit of)* potential function, in advance of polymerization and bio-function, not just after the fact of function *Selection FROM AMONG*.

Even the most extreme reductionism of primitive cellular physiology makes any Metabolism-First model laughable. The needed organization is just too extensive, conceptual, and sophisticated. Even a minimal cell would have been statistically prohibitive as a first-time happenstantial occurrence. The "self-organization" of a protocell violates the Universal Plausibility Principle of science,[80,330] requiring its rejection by peer review for reasons of formal falsification. Its UPM metric of ξ is < 1.

No one has been able to provide a single example of spontaneously generated formal controls from inanimate matter and energy that organizes even the simplest integrated circuit. If ALL biological information in every species arose by "natural process," what exactly is our excuse for not being able to provide a single, simple example of spontaneous P & P independent of investigator involvement in experimental design?

Point mutations are the equivalent of typographical errors. Typographical errors do not improve Ph.D. theses. Typographical errors do not write or improve sophisticated genetic and epigenomic programming, either. The belief that all of genomic programming and precise regulation is the product of physics and chemistry alone is worse than dark-age superstition. Chance and necessity, mass and energy cannot produce *information technology*, as Venter and Dawkins call it.

"Emergence" is what philosophic naturalism appeals to in an attempt to justify belief in the spontaneous appearance of "information technology" and other types of designed and engineered utility. "Emergence" is a wonderfully scientific-sounding term for a completely vacuous concept with

Section 4: Summary

zero empirical support. It also offers no possibility of falsification. This latter point alone eliminates the notion of "emergence" right off from the realm of scientific investigation.

Little needs to be said about emergence, other than the fact that it simply doesn't occur. The only "emergence" ever observed is the self-ordering phenomena of chaos theory. But, self-ordering is dissipative, not sustained. Self-ordering has nothing whatever to do with organization or functionality, and certainly nothing to do with metabolism or life (See Section 3.11).[62,81] Prigogine's "dissipative structures" destroy organization. They never produce it.

Organization is an effect. It must be caused. Since organization is fundamentally formal, it can only be caused by formalistic causation, not by physical "cause and effect."

Limiting causation to the orderliness of nature and to the laws of physics is naïve, at best. It tries to cast a grossly inadequate worldview perimeter around all of the empirical data that must be explained. Way too many puzzle pieces are left over that fail to fit into the supposedly accurate naturalistic picture of reality. The naturalistic metaphysical metanarrative is too small to contain all of the data. Materialism denies the reality of computation and computing. It denies the reality of design and engineering. It ignores the reality of creative art, music, theatre, sport, esthetics and ethics. All of these formalisms require abstract, conceptual, purposeful choice causation, independent of cause-and-effect determinism.

We could try to argue that initial bio-function was exceedingly simple, and gradually found itself in more cooperative schemes. The question then becomes, "How would mere mass/energy interactions have generated *any* formal "cooperative scheme?" Another interesting question might be, "Just how trivial could spontaneously arising function be, and still produce life?" Cooperative subcellular schemes, even in the simplest one-celled life forms known, put to shame the world's finest mainframe computer systems.

Countless astrobiological challenges remain in life-origin science. We continue to deliberately ignore, or sweep under the rug, the single biggest challenge of abiogenesis—the origin of organizing and function-instructing Ontological Prescription and its Processing (P & P). The latter has nothing to do with human epistemological uncertainty, reduced uncertainty, "surprisal," or mutual entropy. P & P precedes all these. P & P not only predates Homo sapiens, it produced human brains and minds.

The only reason this is a problem for naturalistic life-origin science is because of clinging to a starting unproven axiom. In theoretical physics, perfectly pristine mathematical systems can be developed that have no relation or relevance to "the real world." When we realize and finally admit that these mathematical systems "are simply not helpful," or do not correspond to the real world in which we all have to live, we finally wise up and abandon them. We go back and re-examine the starting axiom from which the entire mathematical system was generated. We open-mindedly consider other possible starting axioms. We develop new mathematical systems based on alternative axioms. We then test those mathematical models for correspondence to the repeatedly observed "real world."

Why are we so reluctant to do the same with cosmogony and life-origin science? The answer is that our life-long prized worldview prejudices absolutely forbid it. Our fanatically held articles of faith preclude open-minded consideration of any axiom but the one to which we have always been dogmatically committed. We are locked into our metaphysical imperative, the same as the hierarchy was in Copernicus' day. We deride and mock the hierarchy of Copernicus' day, then turn right around and commit the exact same kind of irrational fanaticism ourselves, denying the plain empirical evidence of the way things actually are. What is the result? A hopeless, on-going, never-ending Kuhnian paradigm rut, one far worse than phlogiston theory or geocentrism ever was.

Section 4: Summary

The only logical escape from having to deal with the necessity of Prescription and its Processing (P & P) would be to 1) redefine life to something different from empirical life, or 2) to argue that initial life was different from current empirical life. In an effort to avoid the nasty problems of needed P & P, both avenues (redefining life, or arguing that initial life was different) have been pursued exhaustively in life-origin literature for many decades.

The first approach involves trying to define down life to the point where it needs no controls, only constraints. Metabolism-first models of life-origin seek to bypass any need for top-down P & P. They argue that initial life was unprescribed, happenstantial, self-organized, self-assembled, spontaneously metabolic, and relentless in its ascent up foothills toward mountain peaks of ever-increasing conceptual complexity. Chance and Necessity are believed to be the sole causative factors. This faith system is utterly bankrupt scientifically. It is nothing more than pure superstition. Worse yet, quality science outrightly refutes it.

The argument that initial life was different is certainly possible. But possibility alone has little to do with empirical science. More often than not, this notion lies closer to Freudian wish-fulfilment in support of one's prior metaphysical commitments, rather than anything scientific. Only one reason exists for dogged pursuit of the possibility that initial life was completely different from the observation of life that science must explain. We are embarrassed by the complete inability of naturalism to explain a single aspect of current life's cybernetic reality.

We finish up where we started. Life origin is *not* about information as we normally think of information. Information is usually *about* reality. Our epistemological concepts of information most often inhabit, or are ascertained by, human knowledge. No knowers, however, let alone humans, existed within the cosmos when life began. *Aboutness* did not exist, and was not the cause of ontological P & P. No searches were being conducted, either. No goals, pursuits or targets existed within a prebiotic environment.

Objective (ontological, not epistemological) Prescription and Processing (P & P) are the issues, independent of knowledge and aboutness. P & P actually produced computation and *caused* formal reality to be introduced into an otherwise purely physical world. This is not just true of life. Inanimate physical interactions are governed by formal computations as well. Mathematics plays the major role in most sciences, physics especially. Most laws are mathematical equations. Constants are numbers. The Formalism > Physicality (F > P) Principle holds true across the board (See Section 3.16). Not even the mathematics of physics, and the scientific method itself, can be stomped into the absurd paradigm rut of naturalism.

Blind belief in an infinite number of universes is pure metaphysics. It is completely outside the bounds of empirical science. Until actually observed, it is nothing more than science fiction. It has no testability. No falsification potential exists. It provides no prediction fulfillments. The non-parsimonious construct of multiverse grossly violates the principle of Ockham's (Occam's) Razor.[533] No logical inference seems apparent to support the strained belief, other than a perceived need to rationalize what we know is statistically prohibitive in the only universe that we *do* experience. Multiverse fantasies tend to constitute a back-door fire escape for when our models hit insurmountable roadblocks in the observable cosmos. When none of the facts fit our favorite model, we conveniently create imaginary extra universes that are more accommodating. This is not science.

Life is an exquisitely programmed constellation of cooperative computations. Programmed computations are effects that can only be *caused* by formal Choice Determinism (CD). Physicodynamic Determinism (PD) (chance and necessity, mass and energy alone) could not possibly have programmed or processed life into existence. The cause of life was Choice-Contingent Causation and Control (CCCC), not some yet-to-be discovered fixed law or spontaneous phase change.

Without acknowledging the objective, ontological reality of Selection *FOR (in pursuit of)* life origin, there will be no solution to the nagging problem of abiogenesis.

Science cannot escape acknowledging the third fundamental category of reality: Choice-Contingency, in addition to Chance-Contingency and Law. Physicodynamics cannot generate mathematical laws, numerical constants, computation, or even statistical analyses. Physicality cannot pursue or produce formal organization, non-trivial utility, holistically integrated metabolic pathways, or life.

Whatever metaphysical worldview rules our puny minds, the facts of ontological *being* prevail. Life could only have arisen from Choice Determinism (CD), not Physicodynamic Determinism (PD). The F > P (Formalism > Physicality) Principle[19,29] applies universally to all life sciences, as it does to physics and chemistry, as the most fundamental principle of science. **See Figure 7, This Section, 4.**

Figure 7. A simple Binary configurable switch.

No physicodynamic force can set this simple binary configurable switch knob SO AS TO favor formal functionality. The switch knob corresponds to Maxwell's demon's trap door. It must be operated by a demon (what philosophers of science call an "agent"). Yes, energy is required to push the knob in one direction or the other. But the choice of *which direction* the knob is pushed IN ORDER TO achieve utility is not determined by physicodynamics.

Conceptual integrated circuits arise only from the far side of The Cybernetic Cut, and enter the physical world only across the one-way Configurable Switch (CS) Bridge from Formalism into Physicality. Universal experience (repeated observation ad infinitum) reveals that if configurable switch-settings contribute to a holistic, non-trivial "usefulness," they were *invariably* set by Choice Determinism (CD), not by Physicodynamic Determinism (PD). Choice-Contingent Causation and Control (CCCC), employing arbitrarily-chosen rules, not laws, is the essence of all formalisms. Formalism is the bottom line of genomically- and epigenomically-programmed, cybernetic life.

Will we ever open our minds to *all* legitimate possibilities as quality scientists? Or will we remain locked forever into our dark dogmatic dungeons of blind faith and superstition in the all-sufficiency of mass and energy?

5. References

1. Turing AM. On computable numbers, with an application to the entscheidungs problem. Proc. Roy. Soc. London Mathematical Society. 1936; 42(Ser 2): 230-265 [correction in 243, 544-546].
2. Abel DL. The biosemiosis of prescriptive information. Semiotica. 2009; 2009(174): 1-19.
3. Hazen RM. The emergence of patterning in life's origin and evolution. Int J Dev Biol. 2009; 53(5-6): 683-692.
4. Hazen RM, Griffin PL, Carothers JM, Szostak JW. Functional information and the emergence of biocomplexity. Proc Natl Acad Sci U S A. May 15 2007; 104 Suppl 1: 8574-8581.
5. Abel DL, ed The First Gene: The Birth of Programming, Messaging and Formal Control. New York, NY: LongView Press-Academic; 2011.
6. Abel DL, Trevors JT. Three subsets of sequence complexity and their relevance to biopolymeric information. Theoretical Biology and Medical Modeling. 2005; 2: 29-45.
7. Abel DL, Trevors JT. More than metaphor: Genomes are objective sign systems. Journal of BioSemiotics. 2006; 1(2): 253-267.
8. Pattee HH. The recognition of description and function in chemical reaction networks. In: Buvet R, Ponnamperuma C, eds. Chemical Evolution and the Origin of Life. Amsterdam: North-Holland; 1971.
9. Pattee HH. Laws and constraints, symbols and languages. In: Waddington CH, ed. Towards a Theoretical Biology. Vol 4. Edinburgh: University of Edinburgh Press; 1972: 248-258.
10. Pattee HH. Universal principles of measurement and language functions in evolving systems. In: Casti JL, Karlqvist A, eds. Complexity, Language, and Life: Mathematical Approaches. Berlin: Springer-Verlag; 1986: 579-581.
11. Pattee HH. The measurement problem in artificial world models. Biosystems. 1989; 23(2-3): 281-289; discussion 290.
12. Pattee HH. The limitations of formal models of measurement, control, and cognition. Applied Mathematics and Computation. 1993; 56: 111-130
13. Pattee HH. Evolving Self-Reference: Matter, Symbols, and Semantic Closure. Communication and Cognition-Artificial Intelligence. 1995; 12: 9-28.
14. Pattee HH. The physics of symbols and the evolution of semiotic controls. In: Coombs Mea, ed. Proc. Workshop on Control Mechanisms for Complex Systems: Addison-Wesley; 1997.
15. Pattee HH. Causation, Control, and the Evolution of Complexity. In: Andersen PB, Emmeche C, Finnemann NO, Christiansen PV, eds. Downward Causation: Minds, Bodies, and Matter. Aarhus, DK: Aarhus University Press; 2000: 63-77.
16. Pattee HH. The physics of symbols: bridging the epistemic cut. Biosystems. Apr-May 2001; 60(1-3): 5-21.

17. Pattee HH. The necessity of biosemiotics: Matter-symbol complementarity. Introduction to Biosemiotics: The New Biological Synthesis. Dordrecht, The Netherlands: Springer; 2007: 115-132.

18. Pattee HH, Kull K. A biosemiotic conversation: Between physics and semiotics. Sign Systems Studies. 2009; 37(1/2).

19. Abel DL. The Formalism > Physicality (F > P) Principle. In: Abel DL, ed. In the First Gene: The birth of Programming, Messaging and Formal Control. New York, New York: Ed. LongView Press-Academic, 2011: Biological Research Division; 2011: 447-492.

20. Abel DL. Prescriptive Information (PI) [Scirus SciTopic Page]. 2009; http://lifeorigin.academia.edu/DrDavidLAbel [Last accessed: January, 2015]; also available from www.researchgate.net

21. Abel DL. What is ProtoBioCybernetics? In: Abel DL, ed. The First Gene: The Birth of Programming, Messaging and Formal Control. New York, N.Y.: LongView Press-Academic: Biolog. Res. Div.; 2011: 1-18.

22. Abel DL. The three fundamental categories of reality. In: Abel DL, ed. The First Gene: The Birth of Programming, Messaging and Formal Control. New York, N.Y.: LongView Press-Academic: Biolog. Res. Div.; 2011: 19-54.

23. D'Onofrio DJ, Abel DL, Johnson DE. Dichotomy in the definition of prescriptive information suggests both prescribed data and prescribed algorithms: biosemiotics applications in genomic systems. Theor Biol Med Model. 2012; 9(1): 8 Open access at http://www.tbiomed.com/content/9/1/8 [Last accessed January, 2015]

24. D'Onofrio DJ, An G. A comparative approach for the investigation of biological information processing: An examination of the structure and function of computer hard drives and DNA. Theoretical Biology and Medical Modeling. 2010; 7(1): 3.

25. MacKay DM. Information, Mechanism and Meaning. Cambridge, MA: M.I.T. Press; 1969.

26. Johnson DE. Programming of Life. Sylacauga, Alabama: Big Mac Publishers; 2010.

27. Johnson DE. 10. What might be a protocell's minimal "genome"? In: Abel DL, ed. The First Gene: The Birth of Programming, Messaging and Formal Control. New York, N.Y.: LongView Press--Academic, Biol. Res. Div.; 2011: 287-303.

28. Johnson DE. Biocybernetics and Biosemiosis. In: Marks II RJ, Behe MJ, Dembski WA, Gordon BL, Sanford JC, eds. Biological Information: New Perspectives. Cornell University Proceedings: World Scientific; 2013: 402-414.

29. Abel DL. Is life unique? Life. 2012; 2(1): 106-134 Open access at http://www.mdpi.com/2075-1729/2072/2071/2106 [Last accessed January, 2015].

30. Abel DL. Moving 'far from equilibrium' in a prebitoic environment: The role of Maxwell's Demon in life origin. In: Seckbach J, Gordon R, eds. Genesis - In the Beginning: Precursors of Life, Chemical Models and Early Biological Evolution. 2012. Dordrecht: Springer; 2012: 219-236.

31. Abel DL. What utility does order, pattern or complexity prescribe? In: Abel DL, ed. The First Gene: The Birth of Programming, Messaging and Formal Control. New York, N.Y.: LongView Press--Academic, Biol. Res. Div.; 2011: 75-116.

32. Abel DL. The Birth of Protocells. In: Abel DL, ed. The First Gene: The Birth of Programming, Messaging and Formal Control. New York, N.Y.: LongView Press--Academic, Biol. Res. Div.; 2011: 189-230.
33. McIntosh AC. Functional Information and Entropy in living systems. In: Brebbia CA, Suchrov LJ, P P, eds. Design and Nature III: Comparing Design in Nature with Science and Engineering. U.K.: WIT Press; 2006.
34. McIntosh AC. Information And Entropy – Top-down Or Bottom-up Development In Living Systems? International Journal of Design & Nature and Ecodynamics. 2010; 4(4): 351-385.
35. Sharov AA. Role of Utility and Inference in the Evolution of Functional Information. Biosemiotics. Apr 1 2009; 2(1): 101-115
36. Sharov AA. Functional Information: Towards Synthesis of Biosemiotics and Cybernetics. Entropy (Basel). Apr 27 2010; 12(5): 1050-1070.
37. Szostak JW. Functional information: Molecular messages. Nature. Jun 12 2003; 423(6941): 689.
38. Teller C, Willner I. Functional nucleic acid nanostructures and DNA machines. Curr Opin Biotechnol. Aug 2010; 21(4): 376-391.
39. Abel DL, Trevors JT. More than Metaphor: Genomes are Objective Sign Systems. In: Barbieri M, ed. BioSemiotic Research Trends. New York: Nova Science Publishers; 2007: 1-15
40. Monod J. Chance and Necessity. New York: Knopf; 1972.
41. Abel DL. The 'Cybernetic Cut': Progressing from Description to Prescription in Systems Theory. The Open Cybernetics and Systemics Journal. 2008; 2: 252-262 Open Access at http://benthamopen.com/contents/pdf/TOCSJ/TOCSJ-2-252.pdf [Last accessed February, 2015].
42. Abel DL. The Cybernetic Cut and Configurable Switch (CS) Bridge. In: Abel DL, ed. The First Gene: The Birth of Programming, Messaging and Formal Control. New York, N.Y.: LongView Press--Academic, Biol. Res. Div.; 2011: 55-74.
43. Rocha LM. Evidence Sets and Contextual Genetic Algorithms: Exploring uncertainty, context, and embodiment in cognitive and biological systems. Binghamton: Systems Science, http: //informatics.indiana.edu/rocha/dissert.html [last accessed January, 2015], State University of New York; 1997.
44. Rocha LM. Syntactic autonomy: or why there is no autonomy without symbols and how self-organizing systems might evolve them. Annals of the New York Academy of Sciences. 2000: 207-223.
45. Rocha LM. Evolution with material symbol systems. Biosystems. 2001; 60: 95-121.
46. Gitt W. In the beginning was information. Master Books 2006.
47. Gitt W, Compton B, Jorge F. Without Excuse. Powder Springs, GA: C. B. Publishers; 2011.
48. Gitt W, Compton R, Fernandez J. Biological Information — What is It? In: Marks II RJ, Behe MJ, Dembski WA, Gordon BL, Sanford JC, eds. Biological Information — New Perspectives2013: 11-25.

Section 5: References

49. Emmeche C, Kull K. Towards a Semiotic Biology: Life is the Action of Signs. Covent Garden, London: Imperial College Press; 2011.
50. Hoffmeyer J, Emmeche C. Code-Duality and the Semiotics of Nature. Journal of Biosemiotics. 2005; 1: 37-91.
51. Kull K. Semiosis includes incompatibility: On the relationship between semiotics and mathematics. In: Bockarova M, Danesi M, Nunez R, eds. Semiotic and Cognitive Science: Essays on the Nature of Mathematics. Muenhen: Lincom Europa; 2012: 330-339.
52. Kull K, Emmeche C, Favareau D. Biosemiotic questions. Biosemiotics. 2008; 1(1): 41-55.
53. Pattee HH. The physics and metaphysics of Biosemiotics. Journal of Biosemiotics. 2005; 1: 303-324.
54. Barbieri M. Has biosemiotics come of age? In: Barbieri M, ed. Introduction to Biosemiotics: The New Biological Synthesis. Dorcrecht, The Netherlands: Springer; 2007: 101-114.
55. Barbieri M. Is the cell a semiotics system? In: Barbieri M, ed. Introduction to Biosemiotics: The New Biological Synthesis. Dordrecht, The Netherlands: Springer; 2007: 179-208.
56. Barbieri M, ed BioSemiotic Research Trends. New York: Nova Science Publishers, Inc.; 2007.
57. Barbieri M. Biosemiotics: a new understanding of life. Naturwissenschaften. 2008; 95: 577-599.
58. Barbieri M. Cosmos and History: Life is semiosis; The biosemiotic view of nature. The Journal of Natural and Social Philosophy. 2008; 4(1-2): 29-51.
59. Rocha LM. The physics and evolution of symbols and codes: reflections on the work of Howard Pattee. Biosystems. Apr-May 2001; 60: 1-4.
60. Rocha LM, Hordijk W. Material representations: from the genetic code to the evolution of cellular automata. Artif Life. Winter-Spring 2005; 11(1-2): 189-214.
61. Abel DL. Constraints vs. Controls: Progressing from description to prescription in systems theory. Open Cybernetics and Systemics Journal. 2010; 4: 14-27 Open Access at: http://benthamopen.com/contents/pdf/TOCSJ/TOCSJ-4-14.pdf [Last accessed: February, 2015].
62. Abel DL, Trevors JT. Self-Organization vs. Self-Ordering events in life-origin models. Physics of Life Reviews. 2006; 3: 211-228.
63. Sproul RC. Not a Chance: the Myth of Chance in Modern Science and Cosmology. Grand Rapids, MI: Baker Books; 1994.
64. Pearle J. Causation. Cambridge: Cambridge University Press; 2000.
65. Resnik MD. Choices: An Introduction to Decision Theory. Minneapolis, Minn: University of Minnesota Press; 1987.
66. Sieb RA. The emergence of consciousness. Med Hypotheses. 2004; 63(5): 900-904.
67. Shannon C. Part I and II: A mathematical theory of communication. The Bell System Technical Journal. 1948; XXVII(3 July): 379-423.

68. Abel DL. The Genetic Selection (GS) Principle [Scirus SciTopic Page]. 2009; http: //lifeorigin.academia.edu/DrDavidLAbel [Last accessed: January, 2015]; also available from www.researchgate.net
69. Overman DL. A Case Against Accident and Self-Organization. New York: Rowman and Littlefield Publishers, Inc.; 1997.
70. Abel DL. The capabilities of chaos and complexity. Society for Chaos Theory: Society for Complexity in Psychology and the Life Sciences; Aug 8-10, 2008; International Conference at Virginia Commonwealth University, Richmond, VA.
71. Abel DL. The GS (Genetic Selection) Principle. Frontiers in Bioscience. 2009; 14(January 1): 2959-2969 Open access at http: //www.bioscience.org/2009/v2914/af/3426/fulltext.htm.
72. Durston KK. Methods: Measuring the 'Functional Sequence Complexity' of proteins: Methods. 2007; http: //www.uoguelph.ca/~kdurston/Methods.rtf.
73. Durston KK, Chiu DKY. Functional Sequence Complexity in Biopolymers. In: Abel DL, ed. The First Gene: The Birth of Programming, Messaging and Formal Control. New York, N.Y.: LongView Press--Academic, Biol. Res. Div.; 2011: 117-133.
74. Durston KK, Chiu DK, Abel DL, Trevors JT. Measuring the functional sequence complexity of proteins. Theoretical biology & medical modelling. 2007; 4: Free on-line access at http: //www.tbiomed.com/content/4/1/47.
75. Szostak JW, Bartel DP, Luisi PL. Synthesizing life. Nature. 2001; 409(3, Jan 18): 387-390.
76. Doya K, Shadlen MN. Decision Making. Current Opinion in Neurobiology. 2012; 22(6): 911–913.
77. Buller DJ, ed Function, Selection and Design. New York: State University of New York Press; 1999. Shaner DE, ed. SUNY Series in Philosophy and Biology.
78. Parmigiani G. Decision Theory: Principles and Approaches (Wiley Series in Probability and Statistics). West Sussex, UK: Wiley; 2009.
79. Abel DL. The capabilities of chaos and complexity. Int. J. Mol. Sci. 2009; 10(Special Issue on Life Origin): 247-291 Open access at http: //mdpi.com/1422-0067/1410/1421/1247 [last accessed: January, 2015].
80. Abel DL. The Universal Plausibility Metric (UPM) & Principle (UPP). Theor Biol Med Model. December 3, 2009 2009; 6(1): 27 Open access at http: //www.tbiomed.com/content/26/21/27 [Last accessed: January 2015].
81. Trevors JT, Abel DL. Chance and necessity do not explain the origin of life. Cell Biol Int. 2004; 28(11): 729-739.
82. Abel DL. Linear Digital Material Symbol Systems (MSS). In: Abel DL, ed. The First Gene: The Birth of Programming, Messaging and Formal Control. New York, N.Y.: LongView Press--Academic, Biol. Res. Div.; 2011: 135-160.
83. Abel DL. The Genetic Selection (GS) Principle. In: Abel DL, ed. The First Gene: The Birth of Programming, Messaging and Formal Control. New York, N.Y.: LongView Press--Academic; 2011: 161-188.

84. Abel DL. Examining specific life-origin models for plausibility. In: Abel DL, ed. The First Gene: The Birth of Programming, Messaging and Formal Control: LongView Press Academic; 2011: 231-272.
85. Abel DL. Life origin: The role of complexity at the edge of chaos. Washington Science 2006; 2006; Headquarters of the National Science Foundation, Arlington, VA.
86. Abel DL. Complexity, self-organization, and emergence at the edge of chaos in life-origin models. Journal of the Washington Academy of Sciences. 2007; 93(4): 1-20.
87. Gilboa I. Decision Theory under Uncertainty. New York: Cambridge University Press; 2009.
88. Bechtel W. Philosophy of Mind: An Overview for Cognitive Science. Hillsdale NJ: Erlbaum; 1988: 44-47.
89. Siewert C. "Consciousness and Intentionality". Stanford Encyclopedia of Philosophy (SEP). Stanford University: Metaphysics Research Lab, CSLI, . .
90. Wilson G, Shpall S. Action. Stanford Encyclopedia of Philosophy 2012.
91. Felsenfeld G. A brief history of epigenetics. Cold Spring Harb Perspect Biol. 2014; 6(1).
92. Jacob P. "Intentionality" Stanford Encyclopedia of Philosophy 2010.
93. Chisholm RM. "Intentionality". The Encyclopedia of Philosophy 4: 201. 1967.
94. Byrne A. "Intentionality." Philosophy of Science: An Encyclopedia. Massachusetts Institute of Technology.
95. Weber M. Indeterminism in Neurobiology. Philosophy of Science. 2007; 72: 663-674.
96. Woodward T, Gills JP. The Mysterious Epigenome : What Lies Beyond DNA. Grand Rapids, MI: Kregel Publications; 2012.
97. Pross A. What is Life? How Chemistry Becomes Biology. Oxford, UK: Oxford University Press; 2012.
98. Orgel LE. The Origins of Life: Molecules and Natural Selection. New York: John Wiley; 1973.
99. Cairns-Smith AG. Seven Clues to the Origin of Life. Canto ed. Cambridge: Cambridge University Press; 1990.
100. Dyson F. Life in the Universe: Is Life Digital or Analog? Paper presented at: NASA Goddard Space Flight Center Colloquiem 1999; Greenbelt, MD.
101. Dyson FJ. Origins of Life. 2nd ed. Cambridge: Cambridge University Press; 1998.
102. Ponnamperuma C. Editorial: space biology and the origin of life. Space Life Sci. 1970; 2(2): 119-120.
103. Ponnamperuma C. Chemical evolution and the origin of life. N Y State J Med. 1970; 70(10): 1169-1174.
104. Ponnamperuma C. Primordial organic chemistry and the origin of life. Q Rev Biophys. 1971; 4(2): 77-106.
105. Ponnamperuma C, K HM, Wickramasinghe N. The physicochemical origins of the genetic code. In: Chela-Flores J, Chadha M, Negron-Mendoza A, Oshima T, eds.

Chemical Evolution: Self-Organization of the Macromolecules of Life. Hampton, VA: Deepak Publishing; 1995: 3-18.

106. Ponnamperuma C, Shimoyama A, Friebele E. Clay and the origin of life. Orig Life. 1982; 12(1): 9-40.

107. Ponnamperuma C, Eirich FR, eds. Prebiological Self-Organization of Matter) Hampton, VA: A. Deepak Publishing; 1990.

108. de Duve C. Blueprint for a Cell: The Nature and Origin of Life. Burlington, NC: Patterson; 1991.

109. Morowitz HJ. Beginnings of cellular life. New Haven: Yale University Press; 1992.

110. Ponnamperuma C, Chela-Flores J, eds. Chemical Evolution: Origin of Life Proceedings of The Trieste Conference on Chemical Evolution and the Origin of Life, 26-30 October, 1992 Hampton, VA A Deepak Publishing; 1993.

111. Kauffman SA. The Origins of Order: Self-Organization and Selection in Evolution. Oxford: Oxford University Press; 1993.

112. Deamer D, Szostak JW, eds. The Origins of Life. Cold Spring Harbor, NY: Cold Spring Harbor Press; 2010. A Cold Spring Harbor Perspectives in Biology Collection.

113. Deamer DW, Szostak JW. The origins of life : a subject collection from Cold Spring Harbor perspectives in biology. Cold Spring Harbor, N.Y.: Cold Spring Harbor Laboratory Press; 2010.

114. Deamer DW, Fleischaker GR. Origins of Life: The Central Concepts Boston, MA: Jones and Bartlett Publishers; 1994.

115. Senapathy P. Independent Birth of Organisms. Madison: Genome Press; 1994.

116. Chela-Flores J, Chadha M, Negron-Mendoza A, Oshima T, eds. Chemical Evolution: Self-organization of the Macromolecules of Life Proceedings of The Trieste Conference on Chemical Evolution and the Origin of Life, 25-29 October 1993 Hampton, VA: A. Deepak Publishing; 1995.

117. Holland JH. Hidden Order: How Adaptation Builds Complexity. Redwood City, CA: Addison-Wesley; 1995.

118. Kauffman S. At Home in the Universe: The Search for the Laws of Self-Organization and Complexity. New York: Oxford University Press; 1995.

119. Kauffman S. Understanding genetic regulatory networks. International Journal of Astrobiology. 2003; 2: 131-139.

120. Kauffman S. Beyond Reductionism: Reinventing the Sacred. Zygon. 2007; 42(4): 903-914.

121. Kauffman S. Question 1: origin of life and the living state. Orig Life Evol Biosph. Oct 2007; 37(4-5): 315-322.

122. Kauffman SA. Investigations. New York: Oxford University Press; 2000.

123. Kauffman SA. On emergence. OLEB. 2010; 40(4-5): 381-383.

124. Kauffman SA. Approaches to the Origin of Life on Earth. Life. 2011; 1(1): 34-48.

125. de Duve C. Vital Dust: The Orign and Evolution of Life on Earth. New York: Basic Books: Harper Collins; 1995.
126. de Duve C. Life Evolving: Molecules, Mind, and Meaning. Oxford: Oxford Univ. Press; 2002.
127. Margulis L, Sagan D. What is Life? London: Weidenfeld and Nicholson; 1995.
128. Margulis L. Symbiosis and evolution. Zurek's Nature. 1971; 225(2): 48-57.
129. Margulis L, Chapman MJ. Endosymbioses: cyclical and permanent in evolution. Trends Microbiol. 1998; 6(9): 342-345; discussion 345-346.
130. Zubay G. Origins of Life on the Earth and in the Cosmos. New York WCB/McGraw Hill; 1996.
131. Chela-Flores J, Raulin F, eds. Chemical Evolution: Physics of the Origins and Evolution of Life, Proceedings of the 4th Trieste Conference on Chemical Evolution, Trieste, Italy, 4-8 September 1995 Netherlands: Kluwer Academic Publishers; 1996.
132. Eigen M. (with Winkler-Oswatitsch, R.), Steps Toward Life: A Perspective on Evolution. Oxford, UK: Oxford University Press; 1992.
133. Eigen M, Schuster P. The Hypercycle: A Principle of Natural Self Organization. Berlin: Springer Verlag; 1979.
134. Lahav N. Biogenesis: Theories of Life's Origin. Oxford: Oxford University Press; 1999.
135. Morowitz HJ. Perspectives on thermodynamics and the origin of life. Adv Biol Med Phys. 1977; 16: 151-163.
136. Morowitz HJ. Phase separation, charge separation and biogenesis. Biosystems. 1981; 14(1): 41-47.
137. Morowitz HJ. Reductionism is not a dirty word. Hosp Pract (Off Ed). 1993; 28(12): 23-24.
138. Morowitz HJ. The theory of biochemical organization, metabolic pathways, and evolution. Complexity. 1999; 4: 39-53.
139. Morowitz HJ. The epistemic paradox of mind and matter. Ann N Y Acad Sci. Apr 2001; 929: 50-54.
140. Morowitz HJ. The Emergence of Everything: How the World Became Complex. New York: Oxford University Press; 2002.
141. Morowitz HJ. The emergence of a new kind of biology. Physical biology. 2014; 11(5): 053001.
142. Morowitz HJ, Heinz B, Deamer DW. The chemical logic of a minimum protocell. Orig Life Evol Biosph. 1988; 18(3): 281-287.
143. Morowitz HJ, Kostelnik JD, Yang J, Cody GD. The origin of intermediary metabolism. Proc Natl Acad Sci U S A. Jul 5 2000; 97(14): 7704-7708.
144. Morowitz HJ, Srinivasan V, Smith E. Ligand field theory and the origin of life as an emergent feature of the periodic table of elements. Biol Bull. Aug 2010; 219(1): 1-6.

145. Brack A. The Molecular Origins of Life: Assembling the Pieces of the Puzzle. New York: Cambridge University Press; 1998.
146. Gross M. Life on the Edge. New York: Plenum Press; 1998.
147. Maynard Smith J. The Problems of Biology. Oxford: Oxford Univ. Press; 1986.
148. Maynard Smith J. The units of selection. Novartis Found Symp. 1998; 213: 203-211.
149. Maynard Smith J. The 1999 Crafoord Prize Lectures. The idea of information in biology. Q Rev Biol. 1999; 74(4): 395-400.
150. Maynard Smith J. The concept of information in biology. Philosophy of Science. 2000; 67 (June): 177-194 (entire issue is an excellent discussion).
151. Maynard Smith J, Szathmary E. The Origins of Life: From Birth of Life to the Origin of Language. Oxford: Oxford University Press; 1999.
152. Maynard-Smith J, Szathmary E. The Major Transitions in Evolution. Oxford: Oxford University Press; 1995.
153. Mayr E. Toward a New Philosophy of Biology. Cambridge: Harvard University Press; 1988.
154. Mayr E. This Is Biology: The Science of the Living World. Cambridge, MA: Harvard University Press; 1997.
155. Mayr E. What Evolution Is. New York: Basic Books; 2001.
156. Mayr E. What Makes Biology Unique? Considerations on the Autonomy of a Scientific Discipline. 2004.
157. Szathmary E. From RNA to language. Curr Biol. 1996; 6(7): 764.
158. Szathmary E. Origins of life. The first two billion years. Nature. 1997; 387(6634): 662-663.
159. Szathmary E. The origin of the genetic code: amino acids as cofactors in an RNA world. Trends Genet. 1999; 15(6): 223-229.
160. Szathmary E. The evolution of replicators. Philos Trans R Soc Lond B Biol Sci. 2000; 355(1403): 1669-1676.
161. Szathmary E. Biological information, kin selection, and evolutionary transitions. Theor Popul Biol. 2001; 59: 11-14.
162. Szathmary E. Why are there four letters in the genetic alphabet? Nature reviews. December 1, 2003 2003; 4(12): 995-1001.
163. Szathmary E. The origin of replicators and reproducers. Philos Trans R Soc Lond B Biol Sci. October 29, 2006 2006; 361(1474): 1761-1776.
164. Szathmary E. Coevolution of metabolic networks and membranes: the scenario of progressive sequestration. Philos Trans R Soc Lond B Biol Sci. Oct 29 2007; 362(1486): 1781-1787.
165. Szathmary E. Evolution. To group or not to group? Science. December 23, 2011 2011; 334(6063): 1648-1649.
166. Szathmary E, Demeter L. Group selection of early replicators and the origin of life. J Theor Biol. 1987; 128(4): 463-486.

167. Szathmary E, Gladkih I. Sub-exponential growth and coexistence of non-enzymatically replicating templates. J Theor Biol. 1989; 138(1): 55-58.

168. Szathmáry E, Jordan F, Pal C. Can genes explain biological complexity? Science. 2001; 292: 1315-1316.

169. Szathmary E, Maynard Smith J. From replicators to reproducers: the first major transitions leading to life. J Theor Biol. 1997; 187(4): 555-571.

170. Szathmary E, Santos M, Fernando C. Evolutionary potential and requirements for minimal protocells. Top Curr Chem. 2005; 259: 167-211.

171. Szathmary E, Smith JM. The evolution of chromosomes. II. Molecular mechanisms. J Theor Biol. 1993; 164(4): 447-454.

172. Szathmary E, Smith JM. The major evolutionary transitions. Nature. 1995; 374(6519): 227-232.

173. Szathmary E, Zintzaras E. A statistical test of hypotheses on the organization and origin of the genetic code. J Mol Evol. 1992; 35(3): 185-189.

174. Ferris JP. The chemistry of life's origin. Chem Eng News. 1984; 62: 22-35.

175. Ferris JP. Prebiotic synthesis: problems and challenges. Cold Spring Harb Symp Quant Biol. 1987; 52: 29-35.

176. Ferris JP. Catalysis and prebiotic RNA synthesis. Orig Life Evol Biosph. 1993; 23(5-6): 307-315.

177. Ferris JP. RNA and the origins of life. Origins of life. 1993; 23(5-6).

178. Ferris JP, ed Origins of Life and Evolution of the Biosphere, Papers presented at the 1996 ISSOL Meeting in Orleans, France. Volume 28, Nos.4-6 October 1998. Boston: Kluwer Academic Publishers; 1998.

179. Ferris JP. Prebiotic synthesis on minerals: bridging the prebiotic and RNA worlds. Biol Bull. 1999; 196(3): 311-314.

180. Ferris JP. Montmorillonite catalysis of 30-50 mer oligonucleotides: laboratory demonstration of potential steps in the origin of the RNA world. Origins of Life and Evolution of the Biosphere. Aug 2002; 32(4): 311-332.

181. Joshi PC, Aldersley MF, Ferris JP. Progress in demonstrating total homochiral selection in montmorillonite-catalyzed RNA synthesis. Biochem Biophys Res Commun. Oct 7 2011; 413(4): 594-598.

182. Joshi PC, Aldersley MF, Price JD, Zagorevski DV, Ferris JP. Progress in studies on the RNA world. Orig Life Evol Biosph. Dec 2011; 41(6): 575-579.

183. Loewenstein WR. The Touchstone of Life: Molecular Information, Cell Communication, and the Foundations of Life. New York: Oxford University Press; 1999.

184. Fry I. The Emergence of Life on Earth: A Historial and Scientific Overview. Piscataway, NJ: Rutgers University Press; 2000.

185. Fry I. ORIGIN OF LIFE: Search for Life's Beginnings. Science. May 26, 2006 2006; 312(5777): 1140-1141.

186. Fry I. The role of natural selection in the origin of life. OLEB. 2011; 41: 3-16.

187. Wills C, Bada J. The Spark of Life: Darwin and the Primeval Soup. Cambridge, MA: Perseus Publishing; 2000.
188. Palyi G, Zucchi C, Caglioti L. Fundamentals of Life. Paris: Elsevier; 2002.
189. Harris LF, Sullivan MR, Hatfield DL. Directed molecular evolution. Orig Life Evol Biosph. 1999; 29(4): 425-435.
190. Harris LF, Sullivan MR, Hickok DF. Conservation of genetic information: a code for site-specific DNA recognition. Proc Natl Acad Sci U S A. 1993; 90(12): 5534-5538.
191. Harris WJ. The origin of life--a master molecule? Adv Sci. 1968; 24(121): 326-332.
192. Harris H. Things Come to Life: Spontaneous Generation Revisited Oxford: Oxford University Press; 2002.
193. Schopt JW, ed Life's Origin: The Beginnings of Biological Evolution Ewing, N. J.: Univer. of California Press; 2002.
194. Fenchel T. Origin and Early Evolution of Life. Oxford: Oxford University Press; 2003.
195. Day W. How Life Began, . Cambridge, MA: Foundation for New Directions; 2002.
196. Hazen RM. Genesis: The Scientific Quest for Life's Origin Washington, D.C. 2005: (Joseph Henry Press; 2005.
197. Hazen RM, Sholl DS. Chiral selection on inorganic crystalline surfaces. Nat Mater. Jun 2003; 2(6): 367-374.
198. Hazen Robert M. Chemical Evolution: An Introduction. Chemical Evolution II: From the Origins of Life to Modern Society. Washington DC: American Chemical Society; 2009: 3-13.
199. Kirschner MW, Gerhart JC. The Plausibility of Life: Resolving Darwin's Dilemma New Haven, Conn: Yale University Press; 2005.
200. Deamer D. First Life. University of California Press; 2011.
201. Jastrow R, Rampino M. Origins of Life in the Universe. New York: Cambridge University Press; 2008.
202. Luisi PL. The Emergence of Life: From chemical Origins to Synthetic Biology. Cambridge: Cambridge Univ. Press; 2006.
203. Luisi PL, Ferri F, Stano P. Approaches to semi-synthetic minimal cells: a review. Naturwissenschaften. 2006; 93: 1-13.
204. Okasha S. Evolution and the Levels of Selection. Oxford: Clarendon Oxford University Press; 2006.
205. Sole R, Goodwin B. Signs of Life: How Complexity Pervades Biology. New York: Basic Books; 2000.
206. Russell M, ed Abiogenesis: How Life Began. The Origins and Search for Life. Cambridge: Cosmology Science Publishers; 2011.
207. Goodsell D. The Machinery of Life. New York, NY: Copernicus: Springer; 2010.
208. Rosen R. Life Itself: A Comprehensive Inquiry into the Nature, Origin, and Fabrication of Life (Complexity in Ecological... Chichester, N.Y.: Columbia University Press; 2005.

Section 5: References

209. Yockey HP. Information Theory, Evolution, and the Origin of Life. Second ed. Cambridge: Cambridge University Press; 2005.
210. Shiller B. Origin of Life: 5th Option. Victoria, BC Canada: Trafford Publishing; 2006.
211. Schopf JW. Life's Origin: the Beginnings of Biological Evolution. Berkeley: University of California Press; 2002.
212. Hodge T. Early Origins of Life: Early Evolution Theory. Kindle Books; 2014.
213. Morowitz HJ. Mayonnaise and the origin of life: Thoughts of Minds and Molecules. Oxford, UK: Ox Bow Pr; 1991.
214. Waddington CH, ed The Origin of Life: Toward a Theoretical Biology, Volume 1. Livingston, N.J.: Aldine Transaction; 2008.
215. Lurquin PF. The Origins of Life and the Universe. New York: Columbia University Press; 2003.
216. Adams FC. Origins of Existence: How Life Emerged in the Universe New York, N.Y.: The Free Press: Simon & Schuster; 2002.
217. Fortey R. Life: A Natural History of the First Four Billion Years of Life on Earth. New York, N.Y.: Vintage Books; 1999.
218. Davies P. The Fifth Miracle: The Search for the Origin and Meaning of Life. Middlesex, U.K.: Penguin Books; 2001.
219. Harold FM. In Search of Cell History: The Evolution of Life's Building Blocks. Chicago, Ill: University of Chicago Press; 2014.
220. Rauchfuss H, Mitchell TN. Chemical Evolution and the Origin of Life. Heidelberg: Springer-Verlog; 2008.
221. Benton MJ, ed The History of Life: A Very Short Introduction. Oxford, UK: Oxford University Press; 2008.
222. Egel R, Lankenau D-H, Mulkidjanian AY, eds. Origins of Life: The Primal Self-Organization. Heidelberg: Springer-Verlog; 2011.
223. Trotman C. The Feathered Onion - Creation of life in the Universe Hoboken, N.J.: John Wiley and Sons; 2004.
224. Luisi PL. The problem of macromolecular sequences: the forgotten stumbling block. Orig Life Evol Biosph. Oct 2007; 37(4-5): 363-365.
225. Luisi PL. Contingency and Determinism in the origin of life, and elsewhere. OLEB. 2010; 40(4-5 October): 356-361.
226. Luisi PL. Open questions on the origins of life: Introduction to the Special Issue. OLEB. 2010; 40: 353-355.
227. Meyer SC. Darwin's Doubt. New York, NY: Harper Collins; 2013.
228. Brillouin L. Life, thermodynamics, and cybernetics. In: Leff HS, Rex AF, eds. Maxwell's Demon, Entropy, Information, and Computing. Princeton: Princeton University Press; 1990.
229. Jukes TH. DDT: Maxwell's Demon. Science. Oct 3 1969; 166(3901): 44.
230. Kieu TD. The second law, Maxwell's demon, and work derivable from quantum heat engines. Phys Rev Lett. Oct 1 2004; 93(14): 140403.

231. Leff HS, Rex AF. Maxwell's Demon, Entropy, Information, Computing. Princeton, N.J.: Princeton Univer. Press; 1990.
232. Maddox J. Maxwell's demon: slamming the door. Nature. Jun 27 2002; 417(6892): 903.
233. McClare CW. Chemical machines, Maxwell's demon and living organisms. Journal of theoretical biology. Jan 1971; 30(1): 1-34.
234. Otsuka J, Nozawa Y. Self-reproducing system can behave as Maxwell's demon: theoretical illustration under prebiotic conditions. Journal of theoretical biology. Sep 21 1998; 194(2): 205-221.
235. Quan HT, Wang YD, Liu YX, Sun CP, Nori F. Maxwell's demon assisted thermodynamic cycle in superconducting quantum circuits. Phys Rev Lett. Nov 3 2006; 97(18): 180402.
236. Shenker O. Maxwell's Demon 2. Entropy, classical and quantum information, computing, by H. Leff and A. Rex. Studies in History and Philosophy of Modern Physics. 2004; 35: 537-540.
237. Stanley M. The pointsman: Maxwell's demon, Victorian free will, and the boundaries of science. J Hist Ideas. Jul 2008; 69(3): 467-491.
238. von Baeyer HC. Maxwell's Demon: Why Warmth Disperses and Time Passes. New York: Random House; 1998.
239. Walker I. Maxwell's demon in biological systems. Acta biotheoretica. 1976; 25(2-3): 103-110.
240. Zaslavsky GM. From Hamiltonian chaos to Maxwell's Demon. Chaos. Dec 1995; 5(4): 653-661.
241. Zurek WH. Algorithmic information content, Church-Turing thesis, and physical entropy and Maxwell's demon. In: Zurek WH, ed. Complexity, Entropy, and the Physics of Information. Redwood City: Addison-Wesley; 1990: 73-89.
242. Hamming RW. Coding and Information Theory Englewood Cliffs, N. J.: Prentice Hall; 1980.
243. Adamala K, Anella F, Wieczorek R, Stano P, Chiarabelli C, Luisi PL. Open questions in origin of life: experimental studies on the origin of nucleic acids and proteins with specific and functional sequences by a chemical synthetic biology approach. Computational and structural biotechnology journal. 2014; 9: e201402004.
244. Kimura M. Neutral Theory of Molecular Evolution. N.Y., N.Y.: Cambridge University Press; 1983.
245. Fontana W, Schuster P. Shaping space: the possible and the attainable in RNA genotype-phenotype mapping. J Theor Biol. 1998; 194(4): 491-515.
246. Viedma C. Formation of peptide bonds from metastable versus crystalline phase: implications for the origin of life. Orig Life Evol Biosph. 2000; 30(6): 549-556.
247. Mojzsis SJ, Arrhenius G, McKeegan KD, Harrison TM, Nutman AP, Friend GRL. Evidence for life on Earth before 3,800 million years ago. Nature. 1996; 384: 55-59.
248. Van Zuilen MA, Lepland A, Arrhenius G. Reassessing the evidence for the earliest traces of life. Nature. Aug 8 2002; 418(6898): 627-630.

249. Freeland SJ, Hurst LD. The genetic code is one in a million. Journal of Molecular Evolution. 1998; 47: 238-248.
250. Becker PB, ENCODEprojectConsortium. A User's Guide to the Encyclopedia of DNA Elements: (ENCODE). PLoS Biology. 2011; 9(4): e1001046.
251. Project E. The ENCODE Project: ENCyclopedia Of DNA Elements: ENCODE. 2012; http: //www.genome.gov/10005107, Last accessed January, 2015.
252. Gerstein MB, Bruce C, Rozowsky JS, et al. What is a gene, post-ENCODE? History and updated definition. Genome Res. June 1, 2007 2007; 17(6): 669-681.
253. Li G-W, Oh E, Weissman JS. The anti-Shine-Dalgarno sequence drives translational pausing and codon choice in bacteria. Nature. 2012; 484(26 APRIL): 538
254. D'onofrio DJ, Abel DL. Redundancy of the genetic code enables translational pausing. Frontiers in Genetics. 2014-May-20 2014; 5: 140 Open access at http: //www.ncbi.nlm.nih.gov/pmc/articles/PMC4033003/.
255. Gesteland RF, Cech TR, Atkins JF. The RNA World. 3 ed. Cold Spring Harbor: Cold Spring Harbor Laboratory Press; 2006.
256. Shapiro R. The improbability of prebiotic nucleic acid synthesis. Orig Life. 1984; 14(1-4): 565-570.
257. Shapiro R. Origins: A Skeptic's Guide to the Creation of Life on Earth. New York: Bantam; 1987.
258. Shapiro R. Prebiotic ribose synthesis: a critical analysis. Orig Life Evol Biosph. 1988; 18(1-2): 71-85.
259. Shapiro R. Prebiotic cytosine synthesis: a critical analysis and implications for the origin of life. Proc Natl Acad Sci U S A. 1999; 96(8): 4396-4401.
260. Shapiro R. A replicator was not involved in the origin of life. IUBMB Life. 2000; 49(3): 173-176.
261. Shapiro R. Comments on `Concentration by Evaporation and the Prebiotic Synthesis of Cytosine'. Origins Life Evol Biosph. 2002; 32(3): 275-278.
262. Shapiro R. Small molecule interactions were central to the origin of life. Quarterly Review of Biology. 2006; 81: 105-125.
263. Shapiro R. A simpler origin of life. Scientific American. 2007; Feb 12.
264. Orgel L. Origin of life. A simpler nucleic acid. Science. Nov 17 2000; 290(5495): 1306-1307.
265. Matray TJ, Gryaznov SM. Synthesis and properties of RNA analogs-oligoribonucleotide N3'-->P5' phosphoramidates. Nucleic Acids Research. 2010; 27(20): 3976-3985.
266. Yu H, Zhang S, Chaput J. Darwinian evolution of an alternative genetic system provides support for TNA as an RNA progenitor. Nat Chem. January 1, 2012 2012; 4(3): 183-187.
267. Kozlov IA, Zielinski M, Allart B, et al. Nonenzymatic template-directed reactions on altritol oligomers, preorganized analogues of oligonucleotides. Chemistry. 2000; 6(1): 151-155.

268. Raine D, Luisi PL. Open Questions on the Origin of Life (OQOL). Orig Life Evol Biosph. Oct 2012; 42(5): 379-383.
269. Alberts B, Bray D, Lewis J, Raff M, Roberts K, Watson JD. Molecular Biology of the Cell. New York: Garland Science; 2002.
270. Edelman GM, Gally JA. Degeneracy and complexity in biological systems. PNAS. 2001; 98: 13763-13768.
271. Mayr E. The place of biology in the sciences and its conceptional structure. In: Mayr E, ed. The Growth of Biological Thought: Diversity, Evolution, and Inheritance. Cambridge, MA: Harvard University Press; 1982: 21-82.
272. Mayr E. Introduction, pp 1-7; Is biology an autonomous science? pp 8-23. In: Mayr E, ed. Toward a New Philosophy of Biology, Part 1. Cambridge, MA: Harvard University Press; 1988.
273. Barbieri M, ed Introduction to Biosemiotics: The New Biological Synthesis. Dordrecht, The Netherlands: Springer-Verlag; 2006.
274. Barbieri M, ed The Codes of Life: The Rules of Macroevolution (Biosemiotics). Dordrecht, The Netherlands: Springer; 2007.
275. Abel DL. Is Life Reducible to Complexity? In: Palyi G, Zucchi C, Caglioti L, eds. Fundamentals of Life. Paris: Elsevier; 2002: 57-72.
276. Bradley D. Informatics. The genome chose its alphabet with care. Science. Sep 13 2002; 297(5588): 1789-1791.
277. Mac Donaill DA. Why nature chose A, C, G and U/T: an error-coding perspective of nucleotide alphabet composition. Orig Life Evol Biosph. Oct 2003; 33(4-5): 433-455.
278. Colaco CA, Macdougall A. Mycobacterial chaperonins: the tail wags the dog. FEMS Microbiol Lett. Jan 2014; 350(1): 20-24.
279. Debès C, Wang M, Caetano-Anollés G, Gräter F. Evolutionary Optimization of Protein Folding. PLoS Comput Biol. 2013; 9(1): e1002861.
280. Dimant H, Ebrahimi-Fakhari D, McLean PJ. Molecular Chaperones and Co-Chaperones in Parkinson Disease. Neuroscientist. Jul 24 2012.
281. Elsasser SJ, D'Arcy S. Towards a mechanism for histone chaperones. Biochim Biophys Acta. Mar 2012; 1819(3-4): 211-221.
282. Dembski W. The Design Inference: Eliminating Chance Through Small Probabilities. Cambridge: Cambridge University Press; 1998.
283. Dembski W. Why evolutionary algorithms cannot generate specified complexity. Paper presented at: Conference on Complexity, Information, and Design: An Appraisal1999; Sante Fe Institute.
284. Dembski W, Kushiner JM, eds. Signs of Intelligence. Grand Rapids, MI: Brazos Press; 2001.
285. Dembski WA. No Free Lunch. New York: Rowman and Littlefield; 2002.
286. Dembski WA. The Design Revolution: Answering the Toughest Questions About Intelligent Design. Downers Grove, IL: Intervarsity Press; 2004.

287. Dembski WA, Wells J. The Design of Life: Discovering Signs of Intelligence in Biological Systems. Dallas: Foundation for Thought and Ethics; 2008.
288. Prigogine I, Stengers I. Order Out of Chaos. London, 285-287, 297-301: Heinemann; 1984.
289. Nicolis G, Prigogine I. Exploring Complexity. New York: Freeman; 1989.
290. Prigogine I. The End of Certainty. New York, 161-162: The Free Press; 1997.
291. Sowerby SJ, Cohn CA, Heckl WM, Holm NG. Differential adsorption of nucleic acid bases: Relevance to the origin of life. Proc Natl Acad Sci U S A. 2001; 98(3): 820-822.
292. Ferris JP, Ertem G. Oligomerization of ribonucleotides on montmorillonite: reaction of the 5'-phosphorimidazolide of adenosine. Science. 1992; 257(5075): 1387-1389.
293. Ferris JP, Hill AR, Jr., Liu R, Orgel LE. Synthesis of long prebiotic oligomers on mineral surfaces. Nature. 1996; 381(6577): 59-61.
294. Poccia N, Ansuini A, Bianconi A. Far from equilibrium percolation, stochastic and shape resonances in the physics of life. Int J Mol Sci. 2011; 12(10): 6810-6833.
295. Tao Y, Jia Y, Dewey TG. Stochastic fluctuations in gene expression far from equilibrium: Omega expansion and linear noise approximation. J Chem Phys. Mar 22 2005; 122(12): 124108.
296. Toussaint O, Remacle J, Dierick JF, et al. Approach of evolutionary theories of ageing, stress, senescence-like phenotypes, calorie restriction and hormesis from the view point of far-from-equilibrium thermodynamics. Mech Ageing Dev. Apr 30 2002; 123(8): 937-946.
297. McIntosh AC. Information and Thermodynamics in Living Systems. In: Marks II RJ, Behe MJ, Dembski WA, Gordon BL, Sanford JC, eds. Biological Information: New Perspectives. Cornell Conference Proceedings: World Scientific; 2013: 179-201.
298. Sewell G. Entropy, Evolution and Open Systems. In: Marks II RJ, Behe MJ, Dembski WA, Gordon BL, Sanford JC, eds. Biological Information: New Perspectives. Cornell University, USA, 31 May – 3 June 2011: : World Scientific; 2011: 168-178.
299. Abel DL. The Cybernetic Cut [Scirus SciTopic Page]. 2008; http://lifeorigin.academia.edu/DrDavidLAbel [Last accessed: January, 2015]; also available from www.researchgate.net
300. Einstein A. Sidelights on Relativity. Mineola, N.Y.: Dover; 1920.
301. Hamming RW. The unreasonable effectiveness of mathematics. The American Mathematical Monthly. 1980; 87(2 February): 81-90.
302. Steiner M. The Applicability of Mathematics as a Philosophical Problem. Cambridge, MA: Harvard University Press; 1998.
303. Wigner EP. The unreasonable effectiveness of mathematics in the natural sciences. Comm. Pure Appl. 1960; 13(Feb).
304. Rocha LM. Selected Self-Organization and the the semiotics of evolutionary systems. In: Salthe S, Van de Vijver G, Delpos M, eds. Evolutionary Systems:

Evolutionary and Biological Perspectives on Selection and Self-Organization: Kluwer Academic publishers; 1998: 341-358.

305. Bohr N. Light and life. Nature. 1933; 131: 421.

306. Rosen M. Fundamentals of Measurement and Representation of Natural Systems. New York: North-Holland; 1978.

307. Liebovitch LS, Tao Y, Todorov AT, Levine L. Is there an error correcting code in the base sequence in DNA? Biophys J. Sep 1996; 71(3): 1539-1544.

308. Church GM, Gao Y, Kosuri S. Next-Generation Digital Information Storage in DNA. Science. August 16, 2012 2012.

309. Freistroffer DV, Kwiatkowski M, Buckingham RH, Ehrenberg M. The accuracy of codon recognition by polypeptide release factors. Proc Natl Acad Sci U S A. Feb 29 2000; 97(5): 2046-2051.

310. Krakauer DC. Darwinian demons, evolutionary complexity, and information maximization. Chaos. Sep 2011; 21(3): 037110.

311. Li Z, Stahl G, Farabaugh PJ. Programmed +1 frameshifting stimulated by complementarity between a downstream mRNA sequence and an error-correcting region of rRNA. Rna. Feb 2001; 7(2): 275-284.

312. Lin L, Hale SP, Schimmel P. Aminoacylation error correction. Nature. 1996; 384(6604): 33-34.

313. Ventegodt S, Dahl Hermansen T, Flensborg-Madsen T, Lyck Nielsen M, Clausen B, Merrick J. Human Development V: Biochemistry Unable to explain the Emergence of Biological Form (morphogenesis), and Therefore a New Principle as Source of Biological Information is Needed. TheScientificWorldJOURNAL. 2006; 6: 1359–1367.

314. Abel DL. The Universal Plausibility Metric (UPM) & Principle (UPP) [Scirus SciTopic Page]. 2010; http: //lifeorigin.academia.edu/DrDavidLAbel [Last accessed: January, 2015]; also available from www.researchgate.net

315. Johnson DE. Notes on "Creation of a Bacterial Cell Controlled by a Chemically Synthesized Genome," Science, 7/2/10, p52-56 http: //www.sciencemag.org/cgi/rapidpdf/science.1190719.pdf 2013; Venter's desription of genomic "software" is available from http: //vimeo.com/21193583; See also: www.scienceintegrity.org/artificial-genome.html.

316. Hawking S. A Brief History of Time. New York: Bantam Books; 1988.

317. Hawking S, Penrose R. The Nature of Space and Time. Princeton, N.J.: Princeton U. Press; 1996.

318. Ross H. Why The Universe Is the Way It Is. Grand Rapids, Michigan: Baker Books; 2008.

319. Ratzsch D. Nature, Design and Science. Albany, N.Y.: State University of New York; 2001.

320. Tarnus R. The Passion of the Western Mind. New York: Ballantine Books; 1991.

321. Hoffmeyer J. Code-duality and the epistemic cut. Ann N Y Acad Sci. 2000; 901: 175-186.

322. von Neumann J, Burks AW. Theory of Self-Reproducing Automata. Urbana,: University of Illinois Press; 1966.
323. Pattee HH. The complementarity principle in biological and social structures. Journal of Social and Biological Structure. 1978; 1: 191-200.
324. Pattee HH. The complementarity principle and the origin of macromolecular information. Biosystems. Aug 1979; 11(2-3): 217-226.
325. Pattee HH. Complementarity vs. reduction as explanation of biological complexity. The American journal of physiology. May 1979; 236(5): R241-246.
326. Pattee HH. Irreducible and complementary semiotic forms. Semiotica. 2001; 134: 341-358.
327. Pattee HH. How does a molecule become a message? In: Lang A, ed. Communication in Development; Twenty-eighth Symposium of the Society of Developmental Biology. New York: Academic Press; 1969: 1-16.
328. Pattee HH. Physical problems of decision-making constraints. Int J Neurosci. Mar 1972; 3(3): 99-106.
329. Pattee HH. Cell psychology: an evolutionary approach to the symbol-matter problem. Cognition and Brain Theory. 1982; 5: 325-341.
330. Abel DL. The Universal Plausibility Metric and Principle. In: Abel DL, ed. The First Gene: The Birth of Programming, Messaging and Formal Control. New York, N.Y.: LongView Press--Academic; 2011: 305-324.
331. Views EN. It Takes Work to Operate a Cell. 2013; July 3, 2013. http://www.evolutionnews.org/2013/07/it_takes_work_t074051.html.
332. Khataee H, Wee-Chung Liew A. A mathematical model describing the mechanical kinetics of kinesin stepping. Bioinformatics. February 1, 2014 2014; 30(3): 353-359.
333. Can S, Dewitt MA, Yildiz A. Bidirectional helical motility of cytoplasmic dynein around microtubules. Elife. 2014; 3: e03205.
334. Cianfrocco MA, Leschziner AE. Traffic control: adaptor proteins guide dynein-cargo takeoff. EMBO J. Jul 24 2014.
335. Cleary FB, Dewitt MA, Bilyard T, et al. Tension on the linker gates the ATP-dependent release of dynein from microtubules. Nat Commun. 2014; 5: 4587.
336. Behe MJ. Darwin's Black Box. New York: Simon & Shuster: The Free Press; 1996.
337. Myer SC. Signature in the Cell. New York, N.Y.: Harper One; 2009.
338. Behe MJ, Dembski W, Meyer SC. Science and Evidence for Design in the Universe. San Francisco, CA: Ignatius Press; 2000.
339. Menuge A. Agents Under Fire: Materialism and the Rationality of Science. Oxford, UK: Rowmand and Littlefield; 2004.
340. Dembski W. Dennett on Competence without Comprehension. 2012; http://www.evolutionnews.org/2012/06/dennett_on_comp061451.html. Accessed June, 2014.
341. Gates B. The Road Ahead. (Revised, 1996) (page 188) ed. London: Penguin; 1996.

342. Adami C. Introduction to Artificial Life. New York: Springer/Telos; 1998.
343. Bruza PD, Song DW, Wong KF. Aboutness from a common sense perspective. JASIS. October 2000; 51(12): 1090-1105.
344. Hjorland B. Towards a theory of aboutness, subject, topicallity, theme, domain, field, content . . ., and relevance. Journal of the American Society of Information Systems and Technology. 2001; 52(9): 774-778.
345. Jablonka E. Information: Its interpretation, its inheritance, and its sharing. Philosophy of Science. 2002; 69: 578-605.
346. Stegmann UE. Genetic Information as Instructional Content. Phil of Sci. 2005; 72: 425-443.
347. Johnson DE. Probability's Nature and Nature's Probability (A call to scientific integrity). Charleston, S.C.: Booksurge Publishing; 2010.
348. Gibson DG, Et al, Venter JC. Creation of a Bacterial Cell Controlled by a Chemically Synthesized Genome. Science Journal. 2010; 329(5987 July 2): 52-56.
349. Venter JC. Venter discusses synthetic life 2010; http: //www.guardian.co.uk/science/video/2010/may/20/craig-venter-new-life-form [Last accessed January, 2015].
350. O'Connell C. Passing the baton of life - from Schrödinger to Venter. New Scientist. 2012; 16(13 July): 14.
351. Monnard PA, Kanavarioti A, Deamer DW. Eutectic phase polymerization of activated ribonucleotide mixtures yields quasi-equimolar incorporation of purine and pyrimidine nucleobases. J Am Chem Soc. Nov 12 2003; 125(45): 13734-13740.
352. Betz K, Malyshev DA, Lavergne T, et al. KlenTaq polymerase replicates unnatural base pairs by inducing a Watson-Crick geometry. Nature chemical biology. 2012; advance online publication.
353. Malyshev DA, Dhami K, Lavergne T, et al. A semi-synthetic organism with an expanded genetic alphabet. Nature. 05/07/online 2014; advance online publication.
354. Dawkins R. The Blind Watchmaker. New York: W. W. Norton and Co.; 1986.
355. Johnson DE. Programming of Life. 2013; http: //www.programmingoflife.info/pol-video.html [Last accessed January, 2015].
356. Tro N. Chemistry in Focus: A Molecular View of Our World. 5th ed. St. Paul, MN: Brooks Cole; 2012.
357. Orgel LE. The origin of life--a review of facts and speculations. Trends Biochem Sci. 1998; 23(12): 491-495.
358. Orgel LE. The implausibility of metabolic cycles on prebiotic earth. PLoS biology. 2008; 6: 5-13.
359. Polanyi M. Life's irreducible structure. Science. 1968 Jun 21 1968; 160(834): 1308-1312.
360. Davies P. How we could create life: The key to existence will be found not in primordial sludge, but in the nanotechnology of the living cell. The Guardian 2002; http: //www.theguardian.com/education/2002/dec/11/highereducation.uk [Last accessed January, 2015].

361. Walker SI, Davies PCW. The algorithmic origins of life. Journal of The Royal Society Interface. February 6, 2013 2013; 10(79).
362. Goldman N, Bertone P, Chen S, et al. Towards practical, high-capacity, low-maintenance information storage in synthesized DNA. Nature. 2013; advance online publication.
363. Dawkins R. Climbing Mount Improbable. New York: W.W. Norton & Co; 1996.
364. Meyer SC. Signature in the Cell. New York: Harper Collins; Reprint edition Harper One (2010); 2009.
365. Forbes N. Imitation of Life: How Biology is Inspiring Computing Cambridge: MIT Press; 2004.
366. Behe MJ. Experimental evolution, loss-of-function mutations, and "the first rule of adaptive evolution". Q Rev Biol. Dec 2010; 85(4): 419-445.
367. Coppedge JF. Evolution: Possible or Impossible? Genes, Proteins, and the Laws of Chance. Grand Rapids, MI: Zondervan; 1973.
368. Corning PA. Control information theory: the 'missing link' in the science of cybernetics. Systems Research and Behavioral Science. 2007; 24(3): 297–311.
369. Corning PA. Evolution 'on purpose': how behaviour has shaped the evolutionary process. Biological Journal of the Linnean Society. 2013: 1-19.
370. Pullen S. Intelligent Design or Evolution? Raleigh, N.C,: Intelligent Design Books; 2005.
371. Broom N. How Blind is the Watchmaker? Downer's Grove, IL: InterVarsity Press; 2001.
372. Thaxton CB, Bradley WL, Olsen RL. The Mystery of Life's Origin: Reassessing Current Theories. Dallas, TX: Lewis and Stanley; 1984.
373. Dembski W. Information as a measure of variation. PCID. 2005; 4.2(November): 1-19. Last accessed Jan 2006 at http: //www.iscid.org/pcid/2005/2004/2002/dembski_information_variation.php.
374. Dembski WA, Ewert W, Marks II RJ. A general theory of information cost incurred by successful search. In: Marks II RJ, Behe MJ, Dembski WA, Gordon BL, Sanford JC, eds. Biological Information: New Perspectives. Cornell University Proceedings: World Scientific; 2013: 26-65.
375. Dembski WA, Marks II RJ. "LIFE'S CONSERVATION LAW: Why Darwinian Evolution Cannot Create Biological Information". In: Gordon B, Dembski WA, eds. The Nature of Nature. Wilmington, Del: ISI Books; 2009.
376. Dembski WA, Marks RJ. Conservation of information in search: Measuring the cost of success. IEEE Transactions on Systems, Man and Cybernetics--Part A: Systems and Humans. 2009; 39(5): 1051-1061.
377. Dembski WA, Marks RJ. Life's Conservation Law: Why Darwinian Evolution Cannot Create Biological Information. In: Gordon B, Dembski WA, eds. The Nature of Nature. Wilmington, Del: ISI Books; 2010.
378. Berlinski D. The Advent of the Algorithm: The Idea that Rules the World. New York: Harcourt, Inc.; 2000.

379. Gryder B, Nelson C, Shepard S. Biosemiotic Entropy of the Genome: Mutations and Epigenetic Imbalances Resulting in Cancer. Entropy. 2013; 15(1): 234-261.
380. Gene M. The Design Matrix: A Consilience of Clues. United States: Arbor Vitae Press; 2007.
381. Sanford JC. Genetic Entropy & The Mystery of the Genome. 2 ed. Lima, N.Y.: Elim Publishing; 2005.
382. Sanford JC. Biological Information and Genetic Theory. In: Marks II RJ, Behe MJ, Dembski WA, Gordon BL, Sanford JC, eds. Biological Information: New Perspectives. Cornell Conference Proceedings: World Scientific 2013: 203-209.
383. Sanford JC, Baumgardner JR, Brewer WH. Selection Threshold Severely Constrains Capture of Beneficial Mutations. In: Marks II RJ, Behe MJ, Dembski WA, Gordon BL, Sanford JC, eds. Biological Information — New Perspectives. Cornell University Proceedings: World Scientific; 2012.
384. Gánti T. Organization of chemical reactions into dividing and metabolizing units: the chemotons. Biosystems. Jul 1975; 7(1): 15-21.
385. Gánti T. A theory of biochemical supersystems and its application to problems of natural and artificial biogenesis. Baltimore: University Park Press; 1979.
386. Gánti T. On the organizational basis of the evolution. Acta Biol. 1980; 31(4): 449-459.
387. Gánti T. Biogenesis itself. J Theor Biol. 1997; 187(4): 583-593.
388. Gánti T. On the early evolutionary origin of biological periodicity. Cell Biol Int. 2002; 26(8): 729-735.
389. Gánti T. Chemoton theory. New York: Kluwer Academic/Plenum Publishers; 2003.
390. Gánti T. The Principles of Life. Oxford, UK: Oxford University Press; 2003.
391. Lee DH, Severin K, Yokobayashi Y, Ghadiri MR. Emergence of symbiosis in peptide self-replication through a hypercyclic network [published erratum appears in Nature 1998 Jul 2; 394(6688): 101]. Nature. 1997; 390(6660): 591-594.
392. Eigen M. Self-organization of matter and the evolution of biological macromolecules. Naturwissenschaften. 1971; 58(In German): 465-523.
393. Eigen M, Biebricher CK, Gebinoga M, Gardiner WC. The hypercycle. Coupling of RNA and protein biosynthesis in the infection cycle of an RNA bacteriophage. Biochemistry. 1991; 30(46): 11005-11018.
394. Eigen M, Gardiner WC, Jr., Schuster P. Hypercycles and compartments. Compartments assists--but do not replace--hypercyclic organization of early genetic information. J Theor Biol. 1980; 85(3): 407-411.
395. Eigen M, Schuster P. The hypercycle. A principle of natural self-organization. Part A: Emergence of the hypercycle. Naturwissenschaften. Nov 1977; 64(11): 541-565.
396. Eigen M, Schuster P. Comments on "growth of a hypercycle" by King (1981). Biosystems. 1981; 13(4): 235.
397. Eigen M, Schuster P, Sigmund K, Wolff R. Elementary step dynamics of catalytic hypercycles. Biosystems. 1980; 13(1-2): 1-22.

398. Segre D, Ben-Eli D, Deamer DW, Lancet D. The lipid world. Orig Life Evol Biosph. 2001; 31(1-2): 119-145.
399. Segre D, Ben-Eli D, Lancet D. Compositional genomes: prebiotic information transfer in mutually catalytic noncovalent assemblies. Proc Natl Acad Sci U S A. 2000; 97(8): 4112-4117.
400. Segre D, Lancet D. Composing Life. EMBO Rep. 2000; 1(3): 217-222.
401. Segre D, Lancet D, Kedem O, Pilpel Y. Graded autocatalysis replication domain (GARD): kinetic analysis of self-replication in mutually catalytic sets. Orig Life Evol Biosph. 1998; 28(4-6): 501-514.
402. Lancet D, Shenhav B. Compositional lipid protocells: reproduction without polynucleotides. In: Rasmussen S, Bedau MA, Chen L, et al., eds. Protocells: Bridging Nonliving and Living Matter. Cambridge, MA: MIT Press; 2009.
403. Shenhav B, Oz A, Lancet D. Coevolution of compositional protocells and their environment. Philos Trans R Soc Lond B Biol Sci. Oct 29 2007; 362(1486): 1813-1819.
404. Shenhav B, Segre D, Lancet D. Mesobiotic emergence: molecular and ensemble complexity in early evolution. Adv. Complex Syst. 2003; 6(15-35).
405. Abel DL. To what degree can we reduce "life" without "loss of life"? In: Palyi G, Caglioti L, Zucchi C, eds. Workshop on Life: a satellite meeting before the Millenial World Meeting of University Professors. Vol Book of Abstracts. Modena, Italy: University of Modena; 2000: 4.
406. Ghose T. Origin of Life Needs a Rethink, Scientists Argue. (LiveScience.com. 2012 (Sunday, December 16,). http: //www.livescience.com/25453-life-origin-reframed.html [Last accessed January, 2015].
407. Nagel T. Mind and Cosmos: Why the Materialist Neo-Darwinian Conception of Nature Is Almost Certainly False. New York, N.Y.: Oxford University Press; 2012.
408. Nagel T. Mind and Cosmos: Why the Materialist Neo-Darwinian Conception of Nature is Almost Certainly False Mind Mind. 2013; 122(486): 582-585.
409. Bonnet J, Subsoontorn P, Endy D. Rewritable digital data storage in live cells via engineered control of recombination directionality. Proceedings of the National Academy of Sciences. May 21, 2012 2012.
410. Kompanichenko V. Origin of life by thermodynamic inversion: a universal process. In: Seckbach J, ed. Genesis-In the Beginning: Precursors of Life, Chemical Models and Early Biological Evolution. London: Springer; 2012.
411. Kompanichenko VN. Distinctive properties of biological systemsL the all-round comparison with other natural systems. Frontier Perspectives. 2003; 12(1): 23-35.
412. Kompanichenko VN. Inversion Concept of the Origin of Life. Orig Life Evol Biosph. May 29 2012.
413. Kompanichenko VN. Thermodynamic inversion and self-reproduction with variations: integrated view on the life-nonlife border. J Biomol Struct Dyn. Feb 2012; 29(4): 637-639.
414. Kompanichenko V. Emergence of biological organization through thermodynamic inversion. Frontiers in Bioscience. 2014; 6: 208-224.

415. Schroedinger E. What is Life? Cambridge: Cambridge University Press; 1955.
416. Rizzotti M, ed Defining Life: The Central Problem in Theoretical Biology. Padova: University of Padova Press; 1996.
417. Palyi G, Zucchi, C.Caglioti. Workshop on Life: a satellite meeting before the Millennial World Meeting of University Professors, 2000; Modena, Italy.
418. Axe DD. Estimating the prevalence of protein sequences adopting functional enzyme folds. J Mol Biol. August 27, 2004 2004; 341(5): 1295-1315.
419. Axe DD. The case against a Darwinian origin of protein folds. BIO-complexity. 2010; 1: 1-12.
420. Ross H. More Than a Theory. Grand Rapids, MI: Baker Books; 2009.
421. Giovannoni SJ, Tripp HJ, Givan S, et al. Genome streamlining in a cosmopolitan oceanic bacterium. Science. Aug 19 2005; 309(5738): 1242-1245.
422. Gilbert JA, Muhling M, Joint I. A rare SAR11 fosmid clone confirming genetic variability in the 'Candidatus Pelagibacter ubique' genome. The ISME journal. Jul 2008; 2(7): 790-793.
423. Meyer MM, Ames TD, Smith DP, et al. Identification of candidate structured RNAs in the marine organism 'Candidatus Pelagibacter ubique'. BMC genomics. 2009; 10: 268.
424. Smith DP, Thrash JC, Nicora CD, et al. Proteomic and transcriptomic analyses of "Candidatus Pelagibacter ubique" describe the first PII-independent response to nitrogen limitation in a free-living Alphaproteobacterium. MBio. 2013; 4(6): e00133-00112.
425. Sowell SM, Norbeck AD, Lipton MS, et al. Proteomic analysis of stationary phase in the marine bacterium "Candidatus Pelagibacter ubique." Appl Environ Microbiol. Jul 2008; 74(13): 4091-4100.
426. Steindler L, Schwalbach MS, Smith DP, Chan F, Giovannoni SJ. Energy starved Candidatus Pelagibacter ubique substitutes light-mediated ATP production for endogenous carbon respiration. PLoS One. 2011; 6(5): e19725.
427. Madrigal AC. To Model the Simplest Microbe in the World, You Need 128 Computers. The Atlantic 2012; http://www.theatlantic.com/technology/archive/2012/07/to-model-the-simplest-microbe-in-the-world-you-need-128-computers/260198/ [Last accessed January, 2015].
428. Karr JR, Sanghvi JC, Macklin DN, et al. A whole-cell computational model predicts phenotype from genotype. Cell. Jul 20 2012; 150(2): 389-401.
429. Grigoryev SA. Nucleosome spacing and chromatin higher-order folding. Nucleus. Nov-Dec 2012; 3(6): 493-499.
430. Wang Y, Xiao X, Zhang J, et al. A complex network of factors with overlapping affinities represses splicing through intronic elements. Nat Struct Mol Biol. 2013; 20(1): 36-45.
431. Norris AD, Calarco JA. Emerging roles of alternative pre-mRNA splicing regulation in neuronal development and function. Frontiers in Neuroscience. 2012-August-21 2012; 6.

432. Hickey SF, Sridhar M, Westermann AJ, et al. Transgene regulation in plants by alternative splicing of a suicide exon. Nucleic Acids Research. February 6, 2012 2012.

433. Barbosa-Morais NL, al e. The Evolutionary Landscape of alternative splicing in vertebrate species. Science. 2012; 388: 1587-1593.

434. Pal S, Gupta R, Kim H, et al. Alternative transcription exceeds alternative splicing in generating the transcriptome diversity of cerebellar development. Genome Res. Aug 2011; 21(8): 1260-1272.

435. Wang RY, Han Y, Krassovsky K, Sheffler W, Tyka M, Baker D. Modeling disordered regions in proteins using Rosetta. PLoS One. 2011; 6(7): e22060.

436. Auer S, Miller MA, Krivov SV, Dobson CM, Karplus M, Vendruscolo M. Importance of metastable states in the free energy landscapes of polypeptide chains. Phys Rev Lett. Oct 26 2007; 99(17): 178104.

437. Yockey HP. Do overlapping genes violate molecular biology and the theory of evolution? J Theor Biol. 1979; 80(1): 21-26.

438. Yockey HP. Rebuttal of "overlapping genes and information theory" [letter]. J Theor Biol. 1981; 91(2): 381-382.

439. Yockey HP. Information Theory and Molecular Biology. Cambridge: Cambridge University Press; 1992.

440. Khabirova E, Moloney A, Marciniak SJ, et al. The TRiC/CCT Chaperone Is Implicated in Alzheimer's Disease Based on Patient GWAS and an RNAi Screen in Abeta-Expressing Caenorhabditis elegans. PLoS One. 2014; 9(7): e102985.

441. Parenti G, Moracci M, Fecarotta S, Andria G. Pharmacological chaperone therapy for lysosomal storage diseases. Future medicinal chemistry. Jun 2014; 6(9): 1031-1045.

442. Taipale M, Tucker G, Peng J, et al. A quantitative chaperone interaction network reveals the architecture of cellular protein homeostasis pathways. Cell. Jul 17 2014; 158(2): 434-448.

443. Weinstock MT, Jacobsen MT, Kay MS. Synthesis and folding of a mirror-image enzyme reveals ambidextrous chaperone activity. Proc Natl Acad Sci U S A. Jul 28 2014.

444. Bell Thomas W, Cline Joseph I. Molecular Machines. Chemical Evolution II: From the Origins of Life to Modern Society. Washington DC: American Chemical Society; 2009: 233-248.

445. Hoffman PM. Life's Ratchet: How Molecular Machines Extract Order from Chaos by Peter Hoffman. Basic Books; 2012.

446. Schneider T. Molecular Information Theory and the Theory of Molecular Machines. 2012; http: //schneider.ncifcrf.gov/ [Last accessed January, 2015] old: http: //alum.mit.edu/www/toms/ [Last accessed January, 2015].

447. Schneider TD. Theory of molecular machines. I. Channel capacity of molecular machines. J Theor Biol. 1991; 148(1): 83-123.

448. Schneider TD. Theory of molecular machines. II. Energy dissipation from molecular machines. J Theor Biol. 1991; 148(1): 125-137.

449. Schneider TD. Sequence logos, machine/channel capacity, Maxwell's demon, and molecular computers; a review of the theory of molecular machines. Nanotechnology. 1994; 5: 1-18.

450. Tang Y, Gao XD, Wang Y, Yuan BF, Feng YQ. Widespread existence of Cytosine methylation in yeast DNA measured by gas chromatography/mass spectrometry. Anal Chem. Aug 21 2012; 84(16): 7249-7255.

451. Kahramanoglou C, Prieto AI, Khedkar S, et al. Genomics of DNA cytosine methylation in Escherichia coli reveals its role in stationary phase transcription. Nat Commun. 2012; 3: 886.

452. Hihath J, Guo S, Zhang P, Tao N. Effects of cytosine methylation on DNA charge transport. J Phys Condens Matter. Apr 25 2012; 24(16): 164204.

453. Craig JM, Wong NC, eds. Epigenetics: A Reference Manual | Book. Norfolk, ENGLAND: Caister Academic Press; 2011.

454. McBrian MA, al e. Histone acetylation regulates intracellular pH. Mol Cell. 2013; 49: 310-321.

455. Turner BM. Histone acetylation and an epigenetic code. Bioessays. 2000; 22(9): 836-845.

456. Barbieri M. The Organic Codes: An Introduction to Semantic Biology. Cambridge: Cambridge University Press; 2003.

457. Barbieri M. Biology with information and meaning. History & Philosophy of the Life Sciences. 2004; 25(2 (June)): 243-254.

458. Bolanos-Garcia VM, Wu Q, Ochi T, Chirgadze DY, Sibanda BL, Blundell TL. Spatial and temporal organization of multi-protein assemblies: achieving sensitive control in information-rich cell-regulatory systems. Philosophical Transactions of the Royal Society A: Mathematical, Physical and Engineering Sciences. June 28, 2012 2012; 370(1969): 3023-3039.

459. Wallace JG, Zhou Z, Breaker RR. OLE RNA protects extremophilic bacteria from alcohol toxicity. Nucleic Acids Res. May 4 2012.

460. von Sternberg R. On the roles of repetitive DNA elements in the context of a unified genomic-epigenetic system. Ann N Y Acad Sci. Dec 2002; 981: 154-188.

461. von Sternberg R, Shapiro JA. How repeated retroelements format genome function. Cytogenetic and genome research. 2005; 110(1-4): 108-116.

462. Cao GS, Liu AL, Li N. [Exploration of the hidden layers of genome.]. Yi chuan = Hereditas / Zhongguo yi chuan xue hui bian ji. Sep 2004; 26(5): 714-720.

463. Dolinoy DC, Weinhouse C, Jones TR, Rozek LS, Jirtle RL. Variable histone modifications at the A(vy) metastable epiallele. Epigenetics. Oct 1 2010; 5(7): 637-644.

464. Micura R, Pils W, Hobartner C, Grubmayr K, Ebert MO, Jaun B. Methylation of the nucleobases in RNA oligonucleotides mediates duplex- hairpin conversion. Nucleic Acids Res. 2001; 29(19): 3997-4005.

465. Duan Z, Andronescu M, Schutz K, et al. A three-dimensional model of the yeast genome. Nature. May 20 2010; 465(7296): 363-367.

466. Tanizawa H, Iwasaki O, Tanaka A, et al. Mapping of long-range associations throughout the fission yeast genome reveals global genome organization linked to transcriptional regulation. Nucleic Acids Res. Dec 2010; 38(22): 8164-8177.

467. Fernandez-Gonzalez R, Ramirez MA, Pericuesta E, Calle A, Gutierrez-Adan A. Histone modifications at the blastocyst Axin1(Fu) locus mark the heritability of in vitro culture-induced epigenetic alterations in mice. Biol Reprod. Nov 2010; 83(5): 720-727.

468. Gonzalez-Lergier J, Broadbelt LJ, Hatzimanikatis V. Theoretical considerations and computational analysis of the complexity in polyketide synthesis pathways. Journal of the American Chemical Society. Jul 13 2005; 127(27): 9930-9938.

469. Robertson M. Gene regulation, evolvability and the limits of genomics. Journal of Biology. 2009; 8(11): 94.

470. Robertson M. The evolution of gene regulation, the RNA universe, and the vexed questions of artefact and noise. BMC Biology. 2010; 8(1): 97.

471. Osawa S. Evolution of the Genetic Code. Oxford: Oxford University Press; 1995.

472. Kay L. Who Wrote the Book of Life? A History of the Genetic Code. Stanford, CA: Stanford University Press; 2000.

473. Vasas V, Fernando C, Santos M, Kauffman S, Szathmary E. Evolution before genes. Biol Direct. January 1, 2012 2012; 7: 1.

474. Carothers JM, Oestreich SC, Davis JH, Szostak JW. Informational complexity and functional activity of RNA structures. J Am Chem Soc. Apr 28 2004; 126(16): 5130-5137.

475. Kim SH, Shin DH, Liu J, et al. Structural genomics of minimal organisms and protein fold space. J Struct Funct Genomics. 2005; 6(2-3): 63-70.

476. Schrum JP, Zhu TF, Szostak JW. The Origins of Cellular Life. Cold Spring Harbor Perspectives in Biology. May 19, 2010; 2(9): a002212

477. Badii R, Politi A. Complexity: hierarchical structures and scaling in physics. Cambridge; New York: Cambridge University Press; 1997.

478. Barham J. A dynamical model of the meaning of information. Biosystems. 1996 1996; 38(2-3): 235-241.

479. Bar-Hillel Y, Carnap R. Semantic Information. British Journal for the Philosophy of Science. 1953; 4: 147-157.

480. Gabora L. Self-Other Organization: Why early life did not evolve through natural selection. J Theor Biol. 2006; 241(3): 443-450.

481. Glass JI, Assad-Garcia N, Alperovich N, et al. Essential genes of a minimal bacterium. PNAS. January 3, 2006 2006: 0510013103.

482. Godfrey-Smith P. Genes do not encode information for phenotypic traits. In: Hitchcock C, ed. Contemporary Debates in Philosophy of Science. London: Blackwell; 2003: 275-289.

483. Griffiths PE. Genetic information: A metaphor in search of a theory. Philosophy of Science. 2001; 68: 394-412.

484. Kupiec J-J. On the lack of specificity of proteins and its consequences for a theory of biological organization. Prog Biophys Mol Biol. 2010; 102(1): 45-52.

485. Lartigue C, Glass JI, Alperovich N, et al. Genome transplantation in bacteria: changing one species to another. Science. Aug 3 2007; 317(5838): 632-638.
486. Smith HO, Hutchison CA, 3rd, Pfannkoch C, Venter JC. Generating a synthetic genome by whole genome assembly: phiX174 bacteriophage from synthetic oligonucleotides. Proc Natl Acad Sci U S A. Dec 23 2003; 100(26): 15440-15445.
487. St. Laurent G, Savva Y, Kapranov P. Dark Matter RNA: an Intelligent Scaffold for the Dynamic Regulation of the Nuclear Information Landscape. Frontiers in Genetics. 2012-April-25 2012; 3.
488. Smith A, Turney P, Ewaschuk R. Self-replicating machines in continuous space with virtual physics. Artif Life. Winter 2003; 9(1): 21-40.
489. Toyabe S, Sagawa T, Ueda M, Muneyuki E, Sano M. Experimental demonstration of information-to-energy conversion and validation of the generalized Jarzynski equality. Nature Physics. 2010; 6 (12), 988-992
490. Van den Broeck C. Thermodynamics of information: Bits for less or more for bits? Nature Physics. 2010; 6, 937-938
491. Djebali S, Lagarde J, Kapranov P, et al. Evidence for transcript networks composed of chimeric RNAs in human cells. PLoS One. 2012; 7(1): e28213.
492. Ye J, Pavlicek A, Lunney EA, Rejto PA, Teng CH. Statistical method on nonrandom clustering with application to somatic mutations in cancer. BMC Bioinformatics. 2010; 11: 11.
493. de Lima RL, Hoper SA, Ghassibe M, et al. Prevalence and nonrandom distribution of exonic mutations in interferon regulatory factor 6 in 307 families with Van der Woude syndrome and 37 families with popliteal pterygium syndrome. Genet Med. Apr 2009; 11(4): 241-247.
494. Tlusty T. A model for the emergence of the genetic code as a transition in a noisy information channel. J Theor Biol. Aug 10 2007.
495. Zhu W, Freeland S. The standard genetic code enhances adaptive evolution of proteins. J Theor Biol. Mar 7 2006; 239(1): 63-70.
496. MacPhee DG, Ambrose M. Spontaneous mutations in bacteria: chance or necessity? Genetica. Jan 1996; 97(1): 87-101.
497. Ratzsch D. TheBattle of Beginnings. Downer's Grove, Illinois: InterVarsity Press; 1996.
498. Ratzsch D. Science and Its Limits. Downers Grove: InterVarsity Press; 2000.
499. Sullivan A. The problem of naturalizing semantics". Language & Communication. 2000; 20(2 April): 179-196.
500. Gonsalvez GB, Long RM. Spatial regulation of translation through RNA localization. F1000 Biol Rep. 2012; 4: 16.
501. Bossi L, Schwartz A, Guillemardet B, Boudvillain M, Figueroa-Bossi N. A role for Rho-dependent polarity in gene regulation by a noncoding small RNA. Genes Dev. Aug 15 2012; 26(16): 1864-1873.
502. Yamaguchi A, Abe M. Regulation of reproductive development by non-coding RNA in Arabidopsis: to flower or not to flower. J Plant Res. Jul 27 2012.

503. Kashida S, Inoue T, Saito H. Three-dimensionally designed protein-responsive RNA devices for cell signaling regulation. Nucleic Acids Res. Jul 18 2012.
504. Kapranov P, St. Laurent G. Dark Matter RNA: Existence, function, and controversy. Frontiers in Genetics. 2012-April-23 2012; 3.
505. Rocha LM. Syntactic Autonomy. Proceedings of the Joint Conference on the Science and Technology of Intelligent Systems 1998; National Institute of Standards and Technology. Gaithersburg, MD: IEEE Press; 1998: 706-711.
506. Beiter T, Reich E, Williams R, Simon P. Antisense transcription: A critical look in both directions. Cellular and Molecular Life Sciences (CMLS). 2009.
507. Dinger ME, Pang KC, Mercer TR, Mattick JS. Differentiating Protein-Coding and Noncoding RNA: Challenges and Ambiguities. PLoS Computational Biology. 11/28 2008; 4(11): e1000176.
508. ENCODE C. An integrated encyclopedia of DNA elements in the human genome. Nature. 2012; 489: 57-74.
509. Encode-Project-Consortium. Identification and analysis of functional elements in 1% of the human genome by the ENCODE pilot project. NATURE. 2007; 447: 799-816.
510. Yelin R, Dahary D, Sorek R, et al. Widespread occurrence of antisense transcription in the human genome. Nat Biotechnol. Apr 2003; 21(4): 379-386.
511. Silvestre DA, Fontanari JF. Package models and the information crisis of prebiotic evolution. J Theor Biol. May 21 2008; 252(2): 326-337.
512. de Duve C. Selection by differential molecular survival: a possible mechanism of early chemical evolution. Proc Natl Acad Sci U S A. 1987; 84(23): 8253-8256.
513. Rosslenbroich B. On the Origin of Autonomy: A New Look at the Major Transitions in Evolution (History, Philosophy and Theory of the Life Sciences). Switzerland: Springer; 2104.
514. Nakabachi A, Yamashita A, Toh H, et al. The 160-kilobase genome of the bacterial endosymbiont Carsonella. Science. Oct 13 2006; 314(5797): 267.
515. Joyce GF, Orgel LE. Prospects for understanding the origin of the RNA World. In: Gesteland RF, Cech TR, Atkins JF, eds. The RNA World. Second ed. Cold Spring Harbor, NY: Cold Spring Harbor Laboratory Press; 1999: 49-78.
516. Kok RA, Taylor JA, Bradley WL. A statistical examination of self-ordering of amino acids in proteins. Origins of life and evolution of the biosphere. 1988; 18(1-2): 135-142.
517. Weiss O, Jimenez-Montano MA, Herzel H. Information content of protein sequences. J Theor Biol. 2000; 206(3): 379-386.
518. ReMine WJ. The Biotic Message. St. Paul, Minnesota: Saint Paul Science; 1993.
519. Roth MJ, Forbes AJ, Boyne MT, 2nd, Kim YB, Robinson DE, Kelleher NL. Precise and parallel characterization of coding polymorphisms, alternative splicing, and modifications in human proteins by mass spectrometry. Mol Cell Proteomics. Jul 2005; 4(7): 1002-1008.
520. Morris SC. Molecules of choice? EMBO Rep. Apr 2012; 13(4): 281.

521. Thieffry D, Sarkar S. Forty years under the Central Dogma. Trends Biochem Sci. 1998; 23: 312-316.

522. Rocha LM. Selected self-organization and the semiotics of evolutionary systems. In: Salthe S, van de Vijver G, Delpos M, eds. Evolutionary Systems: Biological and Epistemological Perspectives on Selection and Self-Organization. The Netherlands: Kluwer; 1998: 341-358.

523. Yockey HP. Origin of life on earth and Shannon's theory of communication. Comput Chem. 2000; 24(1): 105-123.

524. Turvey MT, Kugler PN. A comment on equating information with symbol strings. The American journal of physiology. Jun 1984; 246(6 Pt 2): R925-927.

525. Yockey HP. Informatics, Information Theory, and the Origin of Life. Paper presented at: 4th International Conference on Computational Biology and Genome Informatics2002; Duke University; Research Triangle Park.

526. Küppers B-O. Information and the Origin of Life. Cambridge, MA: MIT Press; 1990.

527. Barrau A. Physics in the multiverse. *CERN Courier* 2007; http://lpsc.in2p3.fr/ams/aurelien/aurelien/CCDecMULTIV.pdf [Last accessed January, 2015]

528. Carr B, ed Universe or Multiverse? Cambridge: Cambridge University Press; 2007.

529. Garriga J, Vilenkin A. Prediction and explanation in the multiverse. Phys.Rev.D 77: 043526,2008. 2008(Subjects: High Energy Physics - Theory (hep-th); Astrophysics (astro-ph); General Relativity and Quantum Cosmology (gr-qc)). arXiv: 0711.2559. Accessed 11/7/2009.

530. Axelsson S. Perspectives on handedness, life and physics. Med Hypotheses. Aug 2003; 61(2): 267-274.

531. Koonin EV. The Biological Big Bang model for the major transitions in evolution. Biol Direct. 2007; 2: 21.

532. Koonin EV. The cosmological model of eternal inflation and the transition from chance to biological evolution in the history of life. Biol Direct. 2007; 2:15.

533. Vitányi PMB, Li M. Minimum Description Length Induction, Bayesianism and Kolmogorov Complexity. IEEE Transactions on Information Theory. 2000; 46(2): 446 - 464.

534. Ashton J. Evolution Impossible. Green Forest, AR: Master Books; 2012.

535. Rana F. The Cell's Design. Grand Rapids, MI: Baker Books; 2008.

536. Rana F, Ross H. Origins of Life. Colorado Springs, Co: NavPress; 2004.

537. Swift D. Evolution Under the Mircrocsope. Sterling University Innovation Park, UK Leighton Academic Press; 2002.

538. Kuhn TS. The Structure of Scientific Revolutions. 2nd 1970 ed. Chicago: The University of Chicago Press; 1970.

539. McGrath A, McGrath JC. The Dawkins Delusion? Downer's Grove, Illinois: IVP Books; 2007.

540. Pattee HH. Artificial Life Needs a Real Epistemology. In: Moran F, ed. Advances in Artificial Life. Berlin: Springer; 1995: 23-38.

541. Ellington AD, Szostak JW. In vitro selection of RNA molecules that bind specific ligands. Nature. 1990; 346(6287): 818-822.

542. Tuerk C, Gold L. Systematic evolution of ligands by exponential enrichment -- RNA ligands to bacteriophage - T4 DNA-polymerase. Science. 1990; 249: 505-510.

543. Robertson DL, Joyce GF. Selection in vitro of an RNA enzyme that specifically cleaves single-stranded DNA. Nature. 1990; 344: 467-468.

544. Abel* DL. Ontolgocial Prescriptive Information (PIo) vs. Epistemological Prescriptive Information (PIe). 2014.

545. Yiping He BV, Victor E. Velculescu, Nickolas Papadopoulos, and Kenneth W. Kinzler. The Antisense Transcriptomes of Human Cells. Science Express. 2008; December 4: 10.1126/science.1163853.

546. Zhang X, Lan W, Ems-McClung SC, Stukenberg PT, Walczak CE. Aurora B phosphorylates multiple sites on mitotic centromere-associated kinesin to spatially and temporally regulate its function. Mol Biol Cell. Sep 2007; 18(9): 3264-3276.

547. Babu MM, Kriwacki RW, Pappu RV. Versatility from Protein Disorder. Science. September 21, 2012 2012; 337(6101): 1460-1461.

548. Wong JT. A co-evolution theory of the genetic code. Proc Natl Acad Sci U S A. 1975; 72(5): 1909-1912.

549. Wong JT. The evolution of a universal genetic code. Proc Natl Acad Sci U S A. 1976; 73(7): 2336-2340.

550. Wong JT. Role of minimization of chemical distances between amino acids in the evolution of the genetic code. Proc Natl Acad Sci U S A. Feb 1980; 77(2): 1083-1086.

551. Wong JT. Evolution of the genetic code. Microbiol Sci. 1988; 5(6): 174-181.

552. Wong JT. Origin of genetically encoded protein synthesis: a model based on selection for RNA peptidation. Orig Life Evol Biosph. 1991; 21(3): 165-176.

553. Wong JT. Coevolution theory of the genetic code at age thirty. Bioessays. Apr 2005; 27(4): 416-425.

554. Wong JT. Question 6: coevolution theory of the genetic code: a proven theory. Orig Life Evol Biosph. Oct 2007; 37(4-5): 403-408.

555. Di Giulio M. The origin of the genetic code: matter of metabolism or physicochemical determinism? J Mol Evol. Oct 2013; 77(4): 131-133.

556. Di Giulio M. The origin of the genetic code in the ocean abysses: new comparisons confirm old observations. J Theor Biol. Sep 21 2013; 333: 109-116.

557. Di Giulio M. A polyphyletic model for the origin of tRNAs has more support than a monophyletic model. J Theor Biol. Feb 7 2013; 318: 124-128.

558. Di Giulio M. The genetic code did not originate from an mRNA codifying polyglycine because the proto-mRNAs already codified for an amino acid number greater than one. J Theor Biol. Sep 9 2014.

559. Di Giulio M, Moracci M, Cobucci-Ponzano B. RNA editing and modifications of RNAs might have favoured the evolution of the triplet genetic code from an ennuplet code. J Theor Biol. Oct 21 2014; 359: 1-5.

560. Di Giulio M. Structuring of the genetic code took place at acidic pH. J Theor Biol. Nov 21 2005; 237(2): 219-226.

561. Di Giulio M. An extension of the coevolution theory of the origin of the genetic code. Biol Direct. 2008; 3: 37.

562. Di Giulio M. Permuted tRNA genes of Cyanidioschyzon merolae, the origin of the tRNA molecule and the root of the Eukarya domain. J Theor Biol. Aug 7 2008; 253(3): 587-592.

563. Di Giulio M. The origin of genes could be polyphyletic. Gene. Dec 15 2008; 426(1-2): 39-46.

564. Di Giulio M. A comparison among the models proposed to explain the origin of the tRNA molecule: A synthesis. J Mol Evol. Jul 2009; 69(1): 1-9.

565. Di Giulio M. A methanogen hosted the origin of the genetic code. J Theor Biol. Sep 7 2009; 260(1): 77-82.

566. Di Giulio M, Amato U. The close relationship between the biosynthetic families of amino acids and the organisation of the genetic code. Gene. Apr 15 2009; 435(1-2): 9-12.

567. Di Giulio M, Kreitman M. Research on the origin of life. Introduction. J Mol Evol. Nov 2009; 69(5): 405.

568. Di Giulio M, Medugno M. The level and landscape of optimization in the origin of the genetic code. J Mol Evol. Apr 2001; 52(4): 372-382.

569. Guimaraes RC. Systemic approaches in genetics. Journal of the Brazilian Assoc. for the Advancement of Science. 1992; 44(5): 314-319.

570. Guimaraes RC. Linguistics of biomolecules and the protein-first hypothesis for the origins of cells. Journal of Biological Physics. 1994; 20: 193-199.

571. Guimaraes RC. Genetic code: dinucleotide type, hydropathic and aminoacyl-t-RNA synthetase class organization. In: Chela-Flores J, Raulin F, eds. Exobiology: Matter, Energy, and Information in the Origin and Evolution of LIfe in the Universe. Netherlands: Kluwer Academic Publishers; 1998: 157-160.

572. Guimaraes RC. Two punctuation systems in the genetic code. In: Chela-Flores J, Owen T, Raulin F, eds. First steps in the origin of life in the universe. Dordrecht, NL: Kluwer Acad. Publ; 2001.

573. Guimaraes RC. The Genetic Code as a Self-Referential and Functional System. International Conference on Computation, Communications and Control Technologies. August 14-17 2004; 7: 160-165.

574. Guimaraes RC. Metabolic basis for the self-referential genetic code. Origins of life and evolution of the biosphere. Nov 6 2011; 41(4): 357-371.

575. Guimarães RC. Formation of the Genetic Code. Update: Nov 2012. Webinar May 2013. 07/2013.

576. Guimaraes RC, Moreira CH, de Farias ST. A self-referential model for the formation of the genetic code. Theory Biosci. Aug 2008; 127(3): 249-270.

577. Wolf YI, Koonin EV. On the origin of the translation system and the genetic code in the RNA world by means of natural selection, exaptation, and subfunctionalization. Biol Direct. 2007; 2(14).

Section 6. Index

A

Abiogenesis, 1-2, 4, 7, 9, 32, 75, 80-82, 85, 93-94, 96, 129-130, 138, 159, 191, 196, 205, 207, 211, 219, 221-222, 247, 250, 253, 261, 264
Abiogenic, 75
Abiotic, 244
Aboutness, 5-6, 151, 230, 262-263
Abstract, 1-5, 7, 16, 20, 25, 28-29, 34-35, 38, 43, 59, 66, 68, 72, 91, 102-103, 107, 117, 122, 125, 132, 146, 161, 187, 193, 201, 222, 244, 257, 260
Acted upon, 4, 58, 126
Adenosine, 74, 134, 223
Agency, 45, 50-55, 137, 185-186, 214-215, 226, 248
Agents, 2, 28, 45, 50-55, 115, 129, 133-134, 213-214, 232, 244
Alan Turing, 1
Algorithms, 13, 19, 39, 77, 87, 90-91, 98, 152, 175, 203, 214, 224, 236, 242
Alphabet, 16, 18, 41-42, 49, 99, 147, 155, 159, 236, 241
Alternatives, 18, 65
Ant eater, 54
Anthropic Principle, 103, 124
Anthropology, 48, 52
Anticipate, 11, 88, 203, 256
Anticipation, 66, 191, 257
Apobetics, 17
Arbitrarily-chosen, 17-18, 36, 50, 117, 145, 266
Arbitrariness, 35-36
Arbitrary, 35, 37, 50, 103, 115, 118, 126, 180, 200, 218, 223, 232
Archeology, 52
Art, 20, 37, 46, 98, 100, 260
Astrobiology, 9
Autocatalytic RNAs, 81
Axiom, 6, 29, 108, 114, 123, 126, 212, 228, 231, 254, 261

B

Bifurcation point, 12, 44, 62, 183
Big Bang, 3, 17, 48, 93, 102, 117, 124, 126
Binary, 12, 15-16, 20-22, 29, 41-42, 44-47, 67, 69, 94-95, 99, 195, 229, 234, 240, 265
Binding energy, 83
Bio-functions, 5, 9, 82, 94, 187, 192, 221
Biochemical pathways, 3, 77, 133, 160, 163, 178, 204, 206, 210, 215, 219, 244, 249
Biological PIo, 94, 197, 212, 215, 219, 232

Bits, 15-16, 36, 45-48, 59-60, 73, 81, 88-89, 109, 111, 165, 192, 234, 236
Brain, 5, 16, 43, 50, 53, 55, 212

C

Catalysis, 7, 75, 79, 81-82, 187, 191, 198
Categorization, 3, 32, 52, 99, 113
Category error, 34, 116, 234
Causal chains, 27, 50
Causation, 1, 3-5, 9, 26-29, 31, 33-36, 38, 40, 47, 50-51, 55, 57, 68-69, 71, 81, 90, 96-97, 99, 104, 107, 115, 120, 127, 141, 150-151, 167, 172, 178, 187, 202, 209, 220-221, 224, 234, 239, 248, 260, 263, 266
Cause-and-effect law of nature, 14
CCCC, 28-29, 31, 33-36, 47, 50-51, 68-69, 71, 81, 90, 94, 96, 99, 104, 107, 109, 120, 151, 167, 183, 209, 212, 220, 222, 239-240, 248, 263, 266
CD, 12-13, 17, 22, 24, 28-29, 31, 33-36, 40, 44, 50-52, 69, 71, 81, 90, 95-99, 103-105, 107, 109, 111, 113, 120, 124, 137, 152, 155, 159-160, 174, 183, 209, 213, 215, 218, 220-222, 224, 228, 233, 239-240, 243, 248, 255, 263-264, 266
Chance and Necessity, 4, 7, 12-13, 28, 30-31, 34-35, 39-40, 43, 46, 51, 53, 55, 58-60, 64, 71, 76, 93, 95-96, 105-106, 113, 121, 124, 127, 129, 148, 152, 155, 160, 162-163, 168, 172, 176, 181, 187, 192-193, 196-197, 203, 206, 209, 217, 220-223, 230, 240, 243, 248, 252, 254, 259, 262-263
Chaos Theory, 2, 23, 29, 37, 68, 96, 98, 100-101, 137, 142, 177, 184, 206, 228, 243, 250, 260
Chaperones, 70, 78, 196
Chicken-and-egg paradox, 85, 226-227
Choice Determinism (CD), 12, 17, 22, 24, 28, 31, 33, 36, 40, 44, 50, 52, 71, 81, 90, 95-99, 103-105, 107, 109, 111, 113, 120, 124, 137, 152, 155, 159-160, 174, 213, 215, 218, 220-222, 239, 243, 255, 263-264, 266
Choice opportunities, 16, 45, 47, 185, 229
Choice-based decisions, 1
Choice-contingency, 2, 11, 20, 24, 26-27, 29, 31-33, 35-36, 40-43, 49-50, 55, 60, 66, 68, 73, 96, 98-99, 104, 112-113, 115, 123, 151-152, 184, 195, 203, 212, 223, 238-240, 248, 258, 264
Choice-Contingent Causation and Control (CCCC), 29, 31, 33-36, 47, 50, 69, 71, 81, 90, 96, 104, 107, 120, 151, 167, 209, 266
Circuit, 19, 21-23, 36, 41-43, 69, 113, 163, 219, 259
Coin flip, 16, 31, 234
Coin flips, 20, 31, 55, 61, 69, 121, 168, 182, 234, 240, 245
Coin tosses, 12, 31
Common-sense language, 21
Communicate, 12, 16, 94, 165, 188
Complexity, 22, 46-47, 49, 52, 73, 82, 90, 104, 110-113, 116, 122-123, 156-157, 162, 179, 192-195, 198, 205, 212, 215, 234, 243, 250, 262
Component parts, 19, 23-24, 28, 32, 41, 72, 77, 131, 163, 213, 239, 241, 257
Computation, 1, 28-29, 36-37, 42, 44, 46, 52, 58-59, 99, 116, 120, 124-125, 131, 148-149, 152, 160-161, 172, 176, 183, 192, 211, 218, 232, 234, 239, 242, 245, 255, 260, 263-264

Computational success, 2, 19, 25, 38, 42, 49, 52, 55, 60, 64-66, 73, 81, 84, 90, 98, 104, 115-116, 120-121, 159-160, 163, 167, 180, 220, 224, 232, 236, 241, 248
Computer programming, 3, 7, 66, 98, 169, 255
Computers, 11, 20, 23, 25, 29-30, 44, 61, 135, 148, 154, 156, 174, 195
Concept, 4-5, 7, 29, 45, 59, 65, 83, 89, 99-100, 107-108, 118, 122, 136, 146, 160, 170, 172, 193-195, 212, 214, 252, 259
Conceptual complexity, 104, 162, 192-193, 195, 262
Conference debates, 9
Configurable Switch Bridge, 40, 43, 241
Configurable Switch Bridge (CS Bridge), 241
Configurable switch-settings, 7, 19, 21-22, 27-28, 64, 66, 74, 77-78, 114, 160, 186, 203, 210, 213, 217, 241, 243, 246, 257, 265
Constraints, 1, 5, 19, 23, 26, 32, 36, 38-40, 42, 55, 60, 63, 74, 83, 96-97, 106, 109, 114, 114-119, 139, 151, 153, 161, 168, 174, 176, 180-181, 186-187, 196, 200, 206-211, 214, 239, 252, 255, 262
Contingency, 2, 11, 20, 23-24, 26-27, 29, 31-33, 35-38, 40-43, 49-50, 55, 60, 62, 66, 68-69, 73, 96, 98-99, 104, 109, 112-113, 115-116, 123, 151-152, 160, 182, 184, 193, 195, 200, 203, 212, 218, 222-223, 227, 233, 238-240, 248, 258, 264
Control mechanisms, 6, 114, 206
Controls, 3-4, 7, 23, 32, 35-36, 38-39, 42-44, 47, 53, 74, 77-78, 80, 84-86, 98, 106-107, 114-116, 118, 121-122, 124, 131, 139, 150-151, 153-155, 162-163, 168, 171, 173-174, 176-177, 180-181, 186-187, 194-196, 198, 200-201, 206, 209-215, 239, 242-243, 249, 252-253, 255, 257-259, 262
Cosmic egg, 93
Cosmogonic, 3
Cosmogony, 4, 48, 261
Cosmology, 4, 48, 124
Crystallization, 23, 112, 203, 250
CS Bridge, 40-44, 108, 128, 241
Cybernetic, 2, 6, 8, 30, 33, 38-44, 52, 57, 60-64, 66, 69, 78, 80, 84, 93, 96, 102-105, 107, 114-115, 119, 121, 124, 127, 129, 141, 145, 150, 152-155, 158-159, 161, 168, 173, 175-176, 186, 197-198, 200, 218, 220-221, 224, 231-232, 240-243, 247, 255, 258, 262, 265-266
Cybernetic Cut, 38-44, 96, 103, 107, 119, 127, 218, 221, 224, 232, 240-243, 265
Cycles, 3, 133, 161, 173-174, 176, 183, 206, 210, 215, 244, 249

D

Decision nodes, 2, 8, 12, 15, 18, 20, 24, 31-33, 35-37, 44, 49, 58, 60, 63-65, 69, 71, 73-74, 78, 83, 90, 96, 98, 115, 150-151, 159-160, 168, 182, 184, 187, 190, 195-196, 205, 208, 212-213, 230, 236-237, 240-241, 245-246, 255
Decision Theory, 12, 33, 37, 46, 62-63, 76, 84, 114, 120-121, 151, 168, 176, 222, 234, 236, 238, 240, 253, 258
Definition of Life, 186
Description, 1, 6, 39, 84, 104, 116, 127, 154-155, 179, 186, 215, 235
Descriptive Information (DI), 84, 129, 151, 215, 233
Desire, 2, 5, 8, 25, 28, 30, 40, 45, 47, 54, 59, 65, 75, 135, 182, 213, 229, 235
Desired destination, 12

Determinism, 3, 11-12, 15, 17, 22, 24, 28-29, 31, 33-36, 38-40, 43-44, 50-52, 58-61, 69, 71, 78, 81, 90, 95-99, 103-105, 107, 109, 111-113, 115, 120-121, 124, 137, 144, 151-152, 155, 159-160, 174, 180, 191, 199-200, 209, 213, 215, 218, 220-222, 239-240, 243, 248, 255, 260, 263-264, 266
Digital, 24, 43, 73, 77, 114, 119, 152-154, 156, 158, 173, 177, 209, 216-218, 223-224, 228-230
Dip-switch settings, 22
Directed evolution, 19, 89-91, 219, 242
Directions, 2, 15, 78, 115, 147, 201, 213
Disobey, 99, 116
Disorder, 103, 110, 112-113, 243, 250
Disorganization, 64, 68, 110, 112-113, 162, 243
Dissipative Structures, 96-98, 100-101, 185, 206, 209, 214, 219, 228, 243, 260
Dissipative structures, 96-98, 100-101, 185, 206, 209, 214, 219, 228, 243, 260
DNA, 9, 46, 61, 77, 88, 147, 149, 151-152, 154-155, 157-158, 164, 167, 170-171, 180, 189, 195, 198-199, 201-202, 208, 213-214, 216, 222, 229, 236-237, 248, 255-257
Double-blind studies, 9, 55, 126
Duplication plus variation, 76, 164-166, 169, 207
Dynamically-inert, 44, 78, 236, 241, 243

E

Efficacious, 1, 8, 32, 36, 47-48, 63, 69-70, 98, 151, 185, 189, 216
Electromagnetic media, 4
Electrons, 18, 184
Emergence, 6, 51-52, 87, 93, 110, 122, 171-172, 177, 180, 184-185, 214, 219, 248, 252-253, 259-260
ENCODE, 85, 95, 104, 160, 188, 236, 251
Encryption/decryption, 74
End-user freedom of choice, 25
Energy, 4-5, 7, 17, 19, 22, 26, 29-31, 35, 40, 46, 53, 59-61, 64, 66, 68-72, 78, 80, 83, 93-94, 96, 101-110, 112-113, 118, 122, 124-126, 129, 131-134, 140, 148-150, 155, 160, 168, 175-176, 180, 188, 192-193, 196, 198, 201, 203, 206, 209-212, 217, 219-223, 230, 239-240, 242, 244, 246, 249-250, 252, 256, 259-260, 263, 265-266
Energy sinks, 78, 196
Engineer, 11, 58-59, 82, 110, 118, 137, 140-141, 155, 226, 235, 245
Entropy, 1, 6-7, 9, 16, 37, 46, 48, 68, 101-102, 104, 106, 110, 112-113, 157, 167, 169-171, 183, 185, 234, 243, 250, 261
Environmentally selectable usefulness, 19
Epigenomic, 3, 71-74, 76-77, 151-152, 159-160, 171, 186, 188, 197-198, 206, 210, 227, 233, 237, 246, 255, 259
Epiphenomenon, 31, 50
Epistemic cut, 119, 127, 240
Epistemological, 5-9, 52, 55, 84, 92-94, 123, 125, 127-128, 210, 215, 226, 229-231, 233, 239, 248-249, 261-263
Epistemology, 5-7, 44, 84, 92, 127, 132, 219, 223, 230, 241, 249, 254

Equipment, 11, 64, 72
Ethics, 37, 46, 98, 100, 260
Eukaryotes, 61, 189
Evolutionary algorithms, 19, 90-91, 203, 236, 242
Excluded middle, 15, 45, 67

F

F > P Principle, 4, 17, 24, 94, 108, 117, 123-126, 150, 232, 253-254
Fallacious, 116
Fallacy, 34, 62, 234
Falsifiability, 9
Far from equilibrium, 44, 67-68, 101-102, 104, 106, 113, 133, 183, 245
Firmware, 21, 65, 86-87, 178
Fitness, 44, 65, 75, 77, 79, 86, 91, 131, 163, 170, 199, 203-204, 209, 214, 219, 227, 246
Fits, 46-47, 49
Fixed Units of measurement, 46, 49
Force constants, 17, 103, 124
Forks in the road, 31, 62, 182, 240, 245
Formal causation, 1, 3, 38, 40
Formalism, 3-5, 17-18, 34-35, 37, 40, 42, 94, 99-100, 105, 108, 111, 113, 117, 123-127, 150, 212, 232, 241, 243, 248, 253-255, 263-266
Formalism > Physicality (F > P) Principle, 212, 253, 263
Freedom from necessity, 60, 195
FSC, 46-47, 49, 111, 234
Function, 1-9, 11-12, 14, 16, 19, 22-23, 25-33, 35, 37, 39-40, 42-48, 51-52, 54-55, 57-59, 61-63, 65-66, 68-75, 77-78, 80-88, 90-91, 94, 96, 98-100, 103, 105-108, 110-111, 113-115, 118-119, 121-123, 127, 129-131, 139, 143, 146, 148, 150, 153, 155, 159-160, 162-164, 168, 171-173, 180-182, 184-186, 188, 191-192, 194, 196-200, 204, 207-208, 212-215, 217, 219-224, 228-230, 232-235, 239, 241-243, 246-247, 250, 252, 258-261
Functional Information (FI), 2, 15, 47, 112, 151, 167, 219, 230, 233-234
Functional Sequence Complexity, 46-47, 49, 111, 234

G

Gaia, 52
Garbage in, Garbage out, 42
Genetic, 9, 39, 43, 57, 65, 70-71, 73, 76-78, 81, 86, 108, 153, 155, 157, 159, 165-166, 170-171, 173-176, 180, 188-189, 191, 197-198, 204, 207-208, 213-219, 224, 241, 246-247, 255, 259
Genetic Selection Principle, 77, 217
Genomic, 3, 53, 71-72, 74, 76-77, 79, 151-152, 159-160, 165, 171, 186, 195, 197, 206-207, 210, 214, 223, 227, 229, 233, 235, 237, 246, 259
Gibbs free energy, 70, 83, 196, 201, 217

Goal, 4-5, 7, 11, 14, 19, 23, 25-28, 30, 32-34, 48, 54, 58-59, 65-67, 72, 84, 89-91, 100, 107, 110, 130, 153, 182, 185-187, 192-193, 195, 204, 208, 215-216, 219, 235, 241-242, 245-246, 248, 251-252
GS Principle, 65, 71, 77-78, 86, 108, 217, 246-247, 253

H

Halting problem, 1, 19, 121, 246
Hamming block code, 108, 246
Hardware, 21, 38, 65, 123, 153, 157, 178, 213
Heat agitation, 8, 27, 38, 58, 74, 105, 116, 202, 220, 240
Heat engine, 66-67, 105, 109-110, 211, 246
Hexanucleotides, 78
Higher level programming, 21, 186
Highly constrained and ordered, 23
Hill, 14, 132-134
Histone proteins, 77, 229
Homeostatic, 30, 77, 91, 170, 188, 216, 245, 247
Homunculus, 43
How-to, 14, 20, 161
Howard H. Pattee, 42
Human knowledge, 1, 5, 7, 47, 92, 127, 130, 210, 226, 229-230, 248, 262
Hurricanes, 96, 98, 203, 228, 244
Hydrotropism, 54
Hypercycles, 164, 176-177, 180-181, 207, 219

I

In advance of any function, 19
Inanimate environments, 28, 30
Inanimate prebiotic environment, 215
Inanimate, prebiotic environment, 110, 128
Inclined plane, 14, 134
Inert and ideal, 67
Information, 1-9, 13-15, 17-18, 26, 39, 47-48, 57-60, 62, 69, 71-74, 76, 84, 93-94, 96, 100, 112-114, 118, 120, 123, 128-129, 142, 144, 149-151, 153-161, 165-181, 184, 186-188, 192, 196-199, 201, 206-207, 209-212, 215-216, 218-222, 229-230, 233-234, 236-237, 239, 241, 247-250, 257-259, 262
Information about, 3, 6, 93, 239
Information potential, 59, 69, 74, 237
Information technology, 153, 155, 158, 161, 171-172, 177-178, 181, 259
Instantiated, 2, 4-5, 16-17, 20-21, 28-29, 33, 42-43, 58, 60, 78, 100, 103, 105, 109, 113-114, 119, 150, 152, 155, 175-176, 186, 193, 214, 216, 220, 241, 246
Instantiation, 11, 20, 22, 28-30, 32, 36, 40-41, 60, 73, 114, 118

Instructions, 1-3, 6, 12-16, 21, 26, 28-29, 32, 49, 53, 57, 61, 64, 70-71, 74, 76, 85, 115, 121, 129, 131, 141-146, 149, 152-154, 161-162, 165-167, 171, 178-180, 186-188, 197-198, 201-203, 207, 210-211, 213, 215-217, 220, 222, 226, 229, 236-237, 241, 247, 255
Integrated circuits, 22-23, 28, 37, 63-64, 77, 114, 163, 227, 265
Integrated function, 3-4, 246
Integration, 19, 23, 32, 41-42, 72, 104, 110, 122, 160, 166, 195, 213, 215, 218-219, 221, 239, 241, 257
Intent, 2, 5, 11, 23, 38-39, 64, 66-67, 105, 115-116, 134, 142, 150, 152, 161-162, 168, 185, 190, 230, 240-241
Intention, 54
Interface, 24, 179

J

Johnson, 155-156, 158, 177

K

Kinesin, 105, 135, 256
Knowers, 5, 262
Knowledge, 1, 5-8, 47, 55, 92-93, 95, 124, 127, 130, 132, 149, 153, 210, 226, 229-230, 241, 248-249, 262-263

L

Language, 3, 15-16, 20-21, 35, 37, 52, 68, 70, 72-73, 98, 100, 108, 118-119, 124-125, 164, 170, 184, 209, 217-218, 222-223, 256
Law-like orderliness, 5, 38, 102, 240
Laws vs. Rules, 115
Learners, 5
Letters to the editor, 9, 55
Life Definition, (See Definition of life), 186
Logic gates, 28, 40-41, 73, 78, 96, 187, 196, 208
Logic theory,, 3, 52, 68, 98, 108, 117, 125, 152, 176, 184
Low-informational, 23, 72, 106

M

Machine, 9, 13-14, 19, 21, 23-24, 42, 70, 80, 105-106, 110, 134-137, 141, 145, 147-149, 156, 174, 199-201, 213, 230, 241, 246

Machinery, 4, 11, 28, 57, 76, 101-102, 140, 169, 174, 221
Macro- evolution, 74
Makes sense, 20
Mass and energy, 4-5, 7, 17, 30, 35, 40, 46, 53, 60, 64, 71, 93, 96, 102, 124-126, 129, 140, 148-149, 155, 160, 176, 180, 192-193, 203, 206, 221-223, 230, 250, 252, 259, 263, 266
Mass/energy interactions, 19, 40, 61, 126, 209, 239, 242, 249, 260
Material medium, 11
Material Symbol Systems (MSS), 16-17, 60
Materialism, 6, 35, 53, 63, 108, 114, 124-126, 164, 232, 249, 251, 254, 260
Mathematics, 3, 17, 35, 37, 52, 68, 98-99, 109, 116-117, 124-125, 152, 164, 175, 184, 209, 228, 251, 263
Matrix, 58, 60, 73, 188
Maxwell's Demon, 21, 37, 41, 66-68, 104-105, 107-109, 122, 133, 211, 245-246, 265
Maze, 31-32, 69
Meaning, 17-18, 37, 46, 50, 103, 109, 120, 122, 132, 148, 157, 171, 184, 189, 215, 218, 225, 234-235, 241
Measurement, 45-49, 126, 233
Metabolism, 2, 7, 30, 72, 77, 84, 91, 94, 105, 110, 113, 122, 129, 133-134, 136, 157, 162-163, 173, 177, 185, 187-188, 192, 199, 204, 206, 210-211, 215-216, 218-219, 221, 244-245, 247, 249, 252-253, 257-260, 262
Metaphysical, 6, 16, 28-29, 35-36, 50-51, 75, 92, 103, 105, 108, 110, 114, 122, 124-125, 138, 144, 152, 172-173, 175, 177, 214, 225-228, 231, 238, 251, 253, 255, 260-262, 264
Metaphysical naturalism, 28, 50-51, 105, 177, 231
Methylation, 74, 77, 163, 187, 196, 198-199, 220, 229, 257
Mind, 5, 16, 18, 31, 36, 43, 50, 52-53, 55, 57, 89, 150, 169, 179, 186-187, 195, 212-213, 220, 223, 251
Mind/body problem, 31, 50, 53
Minimum free-energy sinks, 78
Modification of existing PI, 89, 250
Molecular evolution, 52, 73-75, 77, 79, 81, 84, 86, 89, 100, 102, 130, 143, 203-206, 246-247, 252
Molecular machines, 24, 70, 78, 82, 85, 104-105, 110, 134-137, 141-142, 152, 161, 172, 192, 197, 205, 217, 222, 256
Molecular stability, 75, 79-81, 206
Monod, 151, 224
Motivation, 30, 53-54, 128-129, 131
MSS, 16-17, 19, 60-61, 74, 109, 145, 155, 163, 177, 184, 203, 241, 258
Multi-dimensional, 149, 197, 228, 236, 255, 257
Multi-layered, 149, 197, 228, 236, 255, 257
Mutations, 32-33, 42, 72, 96, 166-167, 169-171, 180, 191, 204, 207, 259
Mutual entropy, 1, 6-7, 9, 16, 46, 48, 157, 167, 234, 261
Mutual- or Self-replication, 81
Mycoplasma, 88, 187, 195, 215, 236

N

Nanocomputers, 135, 137, 152, 211

Natural Selection, 32, 65, 70-72, 74-78, 86, 88, 91, 108, 162, 170, 180, 190, 197, 208, 214, 217, 246-247, 253
Natural selection (NS), 71, 74, 86, 88, 108, 208, 246
Naturalism, 6, 28-29, 50-51, 63, 83, 92, 105, 114, 121, 125-127, 130, 141, 160, 164, 169, 171, 177, 197, 223-224, 227-228, 231-232, 254-255, 259, 262-263
Noise, 59-60, 63, 72, 81, 88-89, 120, 166-169, 202, 207, 222, 256
Non-material Symbol Systems, 15
Non-physical, 1, 18, 38, 40, 43, 68, 127, 251, 255
Non-trivial, 14, 20, 24, 26-27, 29, 42, 51, 55, 57, 59, 66-67, 69, 71, 81, 84-85, 90-91, 96, 98, 105, 109, 113, 117-118, 137-138, 160, 166, 172, 180-181, 186, 190, 219, 221-222, 231, 233, 235, 242-243, 264-265
Nonphysical, 2-4, 7, 16-18, 20, 22, 29, 34-35, 39-41, 52, 98-100, 103, 107, 111, 113, 117, 123-126, 150, 152, 155, 164, 175, 184, 193, 200, 209, 212, 215-216, 220, 228, 232, 241, 255
Nucleic acids, 44
Nucleosides, 7, 73-74, 77, 85, 191, 199

O

Obey rules, 36, 99, 116
Objective reality, 29, 55, 89, 92-93, 114, 145, 214, 221, 227, 240, 247-249, 254-255
Ockham's razor, 64,144
Ockham's (Occam's) razor, 225,263
Ontological being, 6, 29, 55, 92, 132, 255, 264
Ontological Prescription, 6, 9, 95, 101, 123, 155, 249, 261
Open or Close, 12, 107
Operations, 21, 25, 45, 118
Optimization, 11, 19, 39, 42, 91, 135, 137, 141, 172, 192, 194, 219, 241-242, 258
Optimize, 33, 62, 87, 91, 121, 139, 162, 183, 203, 206, 229
Options, 12, 15-16, 18, 27, 31-32, 44-45, 49, 55, 62-63, 65, 69, 74, 90, 109, 113, 150, 195, 229, 236-237, 246, 258
Order, 2, 33, 53-54, 66, 68, 89-90, 94, 97, 99-100, 102-104, 107-108, 110-113, 116, 120, 124, 168, 170, 174, 183, 185, 189, 195, 209, 219, 227, 239, 242-243, 246-247, 250-252, 265
Organization, 2-3, 13, 23-24, 26, 29-33, 37, 42, 51-52, 64, 67-68, 71-72, 78-79, 82, 87, 96-104, 106, 109-113, 116, 121-124, 131, 137, 160-163, 168, 172, 174, 176-178, 181, 183-186, 188, 190, 194, 197, 203-204, 206-211, 214-215, 219-220, 226-228, 231-232, 239, 241-244, 250-253, 255, 257-260, 264
Origin of PIo, 232
Oscillation models of life origin, 182

P

Paper clip, 138-147
Parsimony, 63
Patterns, 51, 112, 123, 252

PD, 12-13, 22, 34-35, 50, 52, 90, 95, 97, 105, 111, 120, 137, 141, 160, 213, 215, 220, 222, 233, 239, 243, 263-264, 266
Phase changes, 20, 24, 35, 59, 160, 184, 230, 243
Philosophic, 6, 28-29, 52, 83, 92, 108, 114, 121-122, 125, 144, 150, 158, 169, 171-172, 196-197, 218, 223-224, 226-228, 251, 253-254, 259
Photons, 17-18
Phototropism, 54
Physical interactions, 5, 27, 35-36, 58, 74, 99-100, 108, 124, 161, 184, 193, 235, 258, 263
Physical laws, 2-3, 18, 61, 73, 114-115, 117-118, 216, 239, 255
Physicalism, 6, 53, 63, 114, 125-126, 164, 175, 228, 232, 241, 254-255
Physicality is sufficient, 6,29,226
Physicality, 2-4, 6-7, 16-18, 20, 22, 28-29, 31, 33-34, 38, 40-43, 90, 94, 103, 105, 107-108, 116-117, 119, 123-127, 132-133, 149-150, 164, 175, 180, 193, 204, 212, 216, 219-220, 226, 232, 235, 239, 241-242, 244-245, 250-251, 253-255, 263-265
Physicochemical, 3, 20, 28, 45, 50, 59, 75, 83, 85, 87, 132, 136, 158, 187, 191-192, 198-199, 213, 218, 230, 244, 248
Physicodynamic Determinism (PD), 12, 35, 50, 52, 90, 97, 105, 111, 120, 160, 215, 220, 222, 239, 243, 263-264, 266
Physicodynamically incoherent, 78
Physicodynamically indeterminate, 22, 220
Physicodynamically-determined, 95, 233
Physicodynamics, 22-23, 38-39, 59, 63, 78, 100, 108, 118, 124, 132, 141, 165, 176, 180, 183-184, 208, 219, 239, 241, 243-244, 264-265
Pi, 2, 4-8, 13-14, 17, 21, 26, 32, 43, 46-48, 50, 57-58, 60-61, 71-72, 76, 89, 100, 113-114, 118, 121, 123, 128-129, 142, 144-145, 148-151, 165-169, 174, 178, 184, 186-188, 190, 197, 199, 201, 207, 209-213, 215, 218, 221-222, 228-230, 233, 236, 241, 247, 250
Place-holders, 16
Plan, 5, 54, 137
Plausibility, 52, 130, 143, 146, 169, 182, 218, 224, 231, 259
Possibility, 48, 62-63, 90, 109, 116, 143, 169, 172, 182, 189, 192, 224, 227, 230, 233-234, 250-251, 260, 262
Post-biotic evolution, 74, 76-77, 84, 88-89, 246-247, 250, 252
Potential binary choice, 45
Pragmatic, 2, 13, 23-25, 27, 36-37, 39, 54, 59, 62, 65-66, 71, 84, 98-99, 107, 110, 113, 116-117, 119, 121, 133, 150, 161, 176, 193, 195, 209, 229, 233, 245, 250
Pragmatically indifferent, 115
Pragmatism, 42, 107, 184
Pre-assumption, 251, 254
Prebiotic environment, 2, 5-9, 30, 51-53, 57, 64, 67-68, 75, 79-80, 82-85, 87-88, 94, 110, 128-131, 133-134, 141, 143-145, 155, 181, 195, 199, 203, 205-206, 208, 211, 214-215, 226, 230, 232, 237, 239, 241, 245, 247, 250, 253, 262
Prediction fulfillments, 9, 51, 138, 167, 177, 207, 225, 235, 243, 263
Prescribed, 1, 3-4, 7-9, 19, 25, 30-31, 33, 46, 55, 62, 64, 70, 72, 78, 80, 86, 91, 98, 105, 122-124, 129, 142-143, 149, 151, 171, 184, 195, 197, 202, 205, 207, 210, 216-217, 222, 226, 229, 253, 256
Prescription, 1-7, 9-15, 17-18, 20, 22, 24-30, 33, 36, 38-40, 44, 46-50, 57, 63-64, 69, 72, 74, 77, 80-82, 84, 87, 90-96, 101, 106-107, 113-115, 123-124, 126-127, 129-130, 143-144, 149,

152, 155, 158-162, 165, 167-168, 171-173, 177, 181, 183, 191, 196, 198, 201-202, 209-213, 215, 217, 219-221, 223-224, 227-230, 232-233, 239, 241, 248-250, 253, 261-263
Prescription of Function (PoF), 7, 28, 228, 250
Prescription Principle, 11, 29, 38, 107, 123-124, 171, 239
Prescription processing, 14
Prescriptive Choice (PC), 44, 46
Prescriptive Information (PI), 2, 6-8, 14, 17, 47-48, 57-58, 60, 71-72, 100, 113, 118, 123, 128, 144, 149-151, 165, 167-168, 174, 184, 187-188, 199, 201, 207, 209, 215, 221, 229, 233, 241, 247, 250
Presupposition, 16, 114, 251, 254
Prigogine, 96-98, 100-101, 185, 214, 243-244, 260
Primordial life, 161, 206, 210, 253
Primordial Prescription, 11, 57
Probability, 7, 26, 33-34, 36, 59, 63-64, 96, 112, 139-140, 143-144, 146-147, 168, 180, 182, 191-192, 211, 258
Processed, 1-2, 4, 6, 18-19, 28, 30, 48, 53, 114, 129-130, 141-142, 147, 161, 195, 211, 230, 263
Processing, 1-6, 10, 13-14, 19-21, 25-26, 28, 30, 46-48, 50, 57, 59, 64-65, 71-72, 76, 80-81, 84-85, 91-95, 101-102, 107, 113-114, 128-130, 142-146, 148, 153, 156-159, 161-162, 172, 174, 177-179, 186-187, 199, 210-212, 218-219, 221, 226-233, 239, 241, 247-249, 253, 255, 261-263
Programming, 2-3, 5-7, 11-12, 15-16, 18-21, 23-25, 27-28, 31, 33, 36, 40, 42-43, 46-49, 53, 59, 61-66, 70-74, 76-80, 83-84, 86-90, 93-96, 98-99, 101, 108-109, 113-117, 119-123, 148-150, 152-153, 155-156, 158-162, 165-169, 175-177, 182-187, 190, 194-197, 199-200, 207-217, 220-223, 226-227, 229-232, 234-237, 243, 245-247, 250, 252, 254-255, 257-259
Programming predetermines, 19
Prokaryotes, 54, 61, 219
Protein folding, 70, 78, 80, 148-149, 155, 197, 200, 202, 229, 256
Proteins, 44, 46, 70, 77-78, 85-86, 88, 105, 134, 136, 142, 153-155, 173, 197, 200-201, 217, 229, 256
ProtoBioCybernetics, 53
Protocells, 6, 129, 189-190, 201, 204, 208, 210, 215, 219, 230, 249
Protometabolism, 3, 8, 72, 85, 133, 185, 204, 218, 244
Psychology, 52
Puffs of smoke, 18
Pure metaphysics, 225, 251, 263
Purpose, 4, 11, 17, 26, 33, 54, 107, 110, 134, 136, 152, 185-186, 241, 251
Purposeful, 1, 5, 12-13, 15-16, 18-24, 26-28, 30-31, 34, 37-38, 41-42, 44-45, 48-51, 53, 57-58, 60-61, 63, 66-69, 74, 81, 83, 85, 90-91, 95, 98-99, 107, 109-110, 113, 120-121, 124, 128, 130, 133, 137, 144, 159, 161, 168, 175, 183-184, 187, 195-196, 212-213, 219, 221, 226, 228, 230, 234, 236, 239-241, 245-246, 248, 252, 255, 260
Purposeful choice, 15-16, 19, 23, 31, 42, 44-45, 49-50, 60, 67-68, 74, 91, 98-99, 107, 221, 239, 260
Pursue, 11, 25, 28, 30, 33, 38, 40, 54, 59-60, 83, 85, 91, 99, 110, 113, 131, 133, 161, 174, 182, 185, 214, 219, 235, 241, 250, 264
Pursuit, 7-8, 25-27, 31-32, 35-37, 49, 53-54, 58-59, 64-67, 69-71, 74, 76-78, 80, 83-84, 86, 88, 90-91, 95, 99, 108, 113, 121, 128, 130, 137, 142, 159, 163-164, 182, 184-185, 190, 192, 195, 209, 214, 217, 220-222, 226, 228, 232, 236, 243, 245-248, 253, 256-257, 259, 262, 264

Q

Quaternary, 15, 74, 78, 190, 229

R

Random, 8, 12, 16, 20, 28, 31-33, 35, 42, 58, 69, 72, 81, 89, 91, 93, 96-97, 111-112, 165, 190-192, 204, 207, 216, 223, 234, 240
Randomness, 12, 52, 69, 71, 90, 97, 110-114, 117, 122, 234, 250
Rats, 31, 69, 240
Reading and executing, 14
Recorded choices, 20
Regulation, 14, 32, 43, 53, 77-78, 84, 95, 114, 122-123, 158, 161, 170, 174, 177, 186-187, 196-198, 207, 210, 212, 215, 217, 226, 233, 237, 252-253, 257-259
Representationalism, 17, 125, 175
Representing, 23, 44-45, 48, 118
Requirements, 11, 29, 81, 94, 137, 147, 189, 195, 201
Ribosome, 70, 80, 85, 142, 148-149, 169, 196, 200, 202, 217
Ribozyme, 7, 80, 85, 191, 213, 235, 242
RNA analogs, 79, 82, 85, 252, 256
Rules are not laws, 14
Rules, not laws, 37, 72, 74, 99, 266

S

Scientific method, 37, 48, 52, 108, 124-125, 152, 164, 169, 175, 228, 232, 251, 263
Scrabble, 17-18, 41, 61
Searches, 6, 93, 95, 128-129, 131, 182, 262
Searches for targets, 6, 93, 128, 131
Second Law, 64, 67-68, 101, 103-104, 106-107, 113-114, 162, 184, 211, 245, 253
Selection FOR, 11, 25, 64-67, 70-71, 74, 76, 78, 80, 82-84, 86, 88, 95, 121, 142, 159, 163, 190, 208-209, 213, 217, 226, 245-248, 253, 256-257, 264
Selection FROM AMONG, 11, 16, 25, 27, 31, 51, 64-65, 70-71, 73-78, 80, 85, 88, 94, 102, 108, 113, 131, 159, 162-164, 190, 209, 213, 216-217, 220, 237, 245-248, 252, 259
Selection Pressure, 71, 75, 79, 86, 88, 204
Selection pressure, 71, 75, 79, 86, 88, 204
Self-awareness, 53
Self-ordering, 23, 29-30, 33, 37, 39, 67, 72, 89, 96-98, 100, 106, 112-113, 177, 181, 203, 209, 214, 227-228, 243, 250, 260
Self-organization, 30, 64, 67, 100, 110, 122, 172, 177, 208, 211, 219, 228, 231, 242-243, 259
Semantic, 15, 59, 132, 151, 157, 229, 243, 250, 252
Semantic Information, 59, 151, 157
Sentience, 2, 132, 244
Sequences, 11, 17, 61, 72-74, 98, 102, 111, 119, 152, 157, 180, 222, 224, 229, 233

Shannon, 7, 15-17, 46-49, 58-60, 69, 72-74, 76, 81, 88-90, 100, 109-112, 120, 151, 157, 165, 167-168, 192, 202, 207, 209, 212, 233-234, 236-237, 241, 250, 256, 258
Shannon combinatorial uncertainty, 16, 89-90, 109, 258
Shannon Uncertainty, 7, 15, 46-47, 69, 73-74, 81, 100, 109-112, 120, 151, 157, 165, 167, 192, 209, 212, 233-234, 236-237, 241, 250, 258
Simple machines, 14, 134
Simplest known cell, 178
Soap bubble-like entities, 6
Software, 18, 21, 25, 31, 48, 62, 65, 86-87, 123, 149-150, 152-154, 157, 178, 245
Spontaneous generation, 51, 83, 130, 138, 140, 142-147, 173, 181, 213, 215, 225, 241
Standard deviation, 33, 59-60
Statistical combinatorialism, 1
Statistical mechanics, 101
Stochastic Theory, 12, 33, 46, 76, 114, 120-121, 151, 157, 233-234, 236, 238, 253, 258
Strings, 21, 68, 97, 114, 182, 192, 196, 199, 201, 216, 218, 222, 230, 243, 252
Subcellular, 24, 37, 44, 46, 53, 63, 76, 84, 87, 104-105, 130-131, 137, 151, 159, 161, 163, 205, 208, 210-211, 218, 220, 225-226, 230, 244, 248-249, 251, 254, 258, 260
Subjective, 6, 8, 37, 92, 94, 132, 219
Superimposed functions, 256
Superimposed layers, 201
Surprisal, 6-7, 261
Survival, 7, 59, 75-79, 86-87, 92, 96, 108, 130, 136, 162, 189, 195, 197, 217, 247, 252
Sustained Functional Systems, 23, 31, 38, 92, 98, 100-101, 104, 133-134, 136-137, 185, 207, 243-244
Symbol syntax, 17, 99
Syntax, 7, 9, 17-18, 46, 73, 80, 86, 88, 99, 145, 155, 175, 191, 196, 208, 213, 217

T

Tabulation of results, 3
Targets, 6-7, 93, 128-131, 230, 262
Thermodynamically open environment, 106
Thermodynamics, 4, 101, 110, 123, 131, 212, 222, 253
Three-dimensional, 7, 18, 24, 44, 70, 105, 149, 163, 194, 196-197, 217, 222, 229
Tinkering, 95
Tokens, 7, 9, 16-18, 41-42, 60-61, 64, 74, 94, 108-109, 150, 155, 164, 176-177, 187, 200, 203, 220, 232, 241, 249, 258
Tornado, 23, 97, 101, 185
TP, 70, 78, 80, 149, 196, 202, 229, 256
Translational Pausing (TP), 70, 78, 80, 149, 196, 202, 229, 256
Transmission theory, 1
Trap door, 21, 37, 41, 66-69, 105, 107-109, 122, 133, 211-212, 240, 245-246, 265
Trial and error searches, 182
Triplet codon, 71, 80, 88, 108, 187, 217-218, 246
Truth, 55, 87
Turing machine, 9, 13-14, 21, 147-149, 200, 230

Turing Tape, 9, 13-14, 18, 21, 147-149, 200, 213, 221, 230, 236

U

Uncertainty, 1, 6-8, 15-16, 27, 29, 32, 36, 44, 46-49, 58-60, 62-63, 69, 73-74, 76, 81, 89-90, 93, 97, 100, 104, 109-113, 120, 151, 157, 165, 167, 183-185, 192, 209, 212, 233-234, 236-237, 241, 249-250, 258, 261
Uncertainty, reduced, 1, 261
Undecidability, 1
Understanding, 6, 55, 92, 94, 132, 150, 167, 170, 174, 198, 212, 226, 232
Units of measure, 46
Universal Contingency Dichotomy (UCD), 239
Universal Determinism Dichotomy (UDD), 34, 120
Universal Selection Dichotomy (USD), 64, 190, 245
Universality of application of theories, laws and paradigms, 9
UPM, 143, 147, 231, 259
UPP, 147, 218
USD, 64, 74-76, 190, 221, 245
Useful Work, 14, 45, 58-60, 68, 80, 107, 110, 132-134, 136, 160, 188, 193, 214, 243-245
Usefulness, 2, 19, 23, 28, 41-42, 54, 58-59, 63, 66, 83, 100, 105, 132, 146, 185, 195, 211, 214, 244, 265
Utility, 2-3, 7-8, 17, 19-20, 23-27, 29-33, 35, 39-40, 42, 49, 51, 54-55, 57-60, 63, 65-66, 68, 70, 74, 84, 87-88, 91, 95, 98-99, 102, 104-105, 107, 111, 113, 115, 128, 132-134, 136-137, 141, 150, 164, 168, 171, 180-183, 192, 194, 203, 205-206, 208, 210, 214, 228, 232, 239, 241-242, 244-247, 250, 252, 259, 264-265

V

Valuation, 54, 129, 131, 214, 232
Value, 1, 8, 28, 34, 46, 49-50, 54, 59, 65, 83, 94, 105, 119, 126, 128, 132-133, 165, 182, 213-214, 231, 241, 244, 258
Venter, 152-156, 158, 177, 259
Voluntary, 36, 53-54, 115

W

Weather front, 24, 98, 137
Weather system, 98
Weighted means, 73, 234
What works best, 111

X

X-axis, 111

Y

Yet-to-be discovered, 63-64, 116, 160, 192, 237, 257-258, 263
Yet-to-be-discovered law, 64
Yogi Berra-ism, 62

Z

Zygote, 43

Made in the USA
Las Vegas, NV
02 April 2025